Tawhid and Shari'ah

Masudul Alam Choudhury

Tawhid and Shari'ah

A Transdisciplinary Methodological Enquiry

palgrave
macmillan

Masudul Alam Choudhury
Postgraduate Program in Islamic Economics & Finance,
Faculty of Economics (IEF)
Trisakti University
Jakarta, Indonesia

ISBN 978-3-030-49086-7 ISBN 978-3-030-49087-4 (eBook)
https://doi.org/10.1007/978-3-030-49087-4

This Palgrave Macmillan imprint is published by the registered company Springer Nature Switzerland AG.
The registered company address is: Gewerbestrasse 11, 6330 Cham, Switzerland

Qur'an *53:1–5*

By the star when it descends,
Your companion [Muhammad] has not strayed, nor has he erred,
Nor does he speak from [his own] inclination.
It is not but a revelation revealed,
Taught to him by one intense in strength—

Ahadith Al-Bukhari

When I love him I become his hearing with which he hears, his seeing with which he sees, his hand with which he strikes, and his leg with which he walks; and if he asks (something) from Me, I give him, and if he asks My Protection (refuge), I protect him.

Foreword

It gives me a great pleasure to write a few words about the book entitled *Tawhid and Shari'ah* (A transdisciplinary methodological Enquiry), written by Prof Masudul Alam Choudhury. The author has already written a number of books related to the core principles of Islam. The present one is a useful addition and is interesting from many points of view.

The central theme of this work is the expounding of the view held by the author that the scholars and their sectarian schools along with the present days' area of 'shari'ah compliance' in legal legitimation and product development have totally forgotten the explanatory realm of the greatest worldview of all in the Qur'an and the sunnah. This is Tawhid understood as 'law' and bestowed by its overarching methodological worldview that applies to the generality and particulars of 'everything'.

At this point, I feel tempted to make a scientific worldview of these two words 'law' and 'everything' from the Qur'anic point of view. In the Qur'an, two kinds of laws are mentioned. The first kind of laws refer to the governance of the heavens of the earth and all that has been created in Nature {7:85': Do they see nothing in the government of the heavens and the earth and all that Allah has created?….}.The second kind of laws refer to those relating to the ways of living of mankind which have regard for other forms of life and environment. The first kind of laws (gravitation force law, the electromagnetic force law, the strong nuclear force law and the weak nuclear force law) have been discovered by scientists. These laws created after the Big Bang are responsible for the sustenance of living and innate matter ranging from the biggest of the big to the smallest of the small and bear witness to Tawhid: There is only one Creator, one

Law-giver, one Sustainer and Cherisher of the Worlds. There is one more thing: the laws are not laws per se, there is an element of Mercy and consideration for life to thrive up to a certain point of time.

The second kind of laws refer to the social laws of living of the members of mankind as a community as Vice-Gerents of God on earth. Both these two kinds of laws are not meant for a section of mankind, they are meant for all members of mankind. The Qur'an has been mentioned as not only a code of living for Muslims but the whole of mankind (Hudallil Naas). Also the last prophet of Islam (Hazrat Mohammad (PBUH)) has been mentioned in the Qur'an not as a mercy for Muslims alone but for the whole of Creation.

It is interesting to mention here that these two kinds of laws have a universal character and all-encompassing worldview which must not be lost sight of while dealing with Tawhid-based Shari'ah made applicable to Muslims with the second class of laws. In reality, the applicability was meant for all. However, there is no compulsion in religion as emphasized in the Qur'an; God has given man a set of social laws from time to time for a good living on earth. To Mohammad (PBUH) came the final laws. It is up to men and women to follow them or not; they would be accountable to the Lord of the Universe on the final day of judgement. That is, the part of Tawhid which has to look after the worldly requirements of all beings, I repeat all beings, including their economic wellbeing.

Next, I come to the worldview of Tawhid relating to the generality and particulars of 'everything'. It has been mentioned in the Qur'an (2:29) that "it is He who has created for you (meaning mankind) 'all things' that are on earth". This verse has obviously a direct reference to 'every' resource for 'everyone'. This verse has also a deep implication for the distribution of everything (as a resource) to all members of the human family. Again, for one who cares for a Tawhid-based world economic system, reference has to be made to the Qur'anic Verse 59:7, which asserts that wealth should not be a circuitry amongst the rich; unfortunately, this phenomenon has become an order of the day within societies, and between societies, much against Tawhid and Shari'ah. A heed has to be taken also of Verse 51:19, which indicates that on the Day of Judgement it would be mentioned with appreciation that in their (Righteous ones') wealth and possessions (was remembered) the right of the (needy), him who asked, and him who (for some reason) was prevented (from asking). All these do point out that Tawhid-based Shari'ah has an infinite potential for the growth of a just worldwide economic system. For quite sometime the economic order of the world has been revisited from time to time. The world witnessed the North South Dialogue, North-South and South-South Co-operation, the New International Economic Order, Strategic Sustainable Development Goals, and so on. Poverty still remains rampant. Many believed that the New Digital Order would reduce the gap between the rich and the poor; in practice, this did not happen; the gap widened.

In the light of what has been elaborated so far, it must be said that Professor Choudhury has very rightly propounded in his present book the idea namely that the primordial ontological premise of Islamic monotheistic worldview in the scheme and order of 'everything' was never invoked to explain its unfathomed capability, that is, to interrelate the organic monotheistic unity of knowledge with the generalities and particularities of the world-system along with all the ensuing details. The author also holds the view that the clamour around Shari'ah in respect of worldly matters (muamalat) now remains an empty pursuit of mind and matter of a disengaged Muslim society. The present work is a serious critique of the School of Shari'ah in the face of the indelible Tawhid as law in respect of its methodological worldview of unity of knowledge (monotheism) as the primal ontology of 'everything' that thereby emanates.

In the end, I would like to emphasize that although the principal objective of the work is to invoke the nature and logic of the true Islamic law of Tawhid and its ontological and functional relationship with the generalities and specifics of all world-systems, the author would do well to illuminate the readers with further light on the pervasiveness of application of Tawhid in certain areas from the domain of 'everything', for example the burning issues of the day, namely world economic system, man-made global changes, human rights and obligations, a just and fair system of global justice, and so on. This would be especially suitable to further illustrate the point that Tawhid has an inherent capacity of thinking about all humans and of dealing with all humans, keeping in view the Qur'anic assertion that mankind was just one single nation (Qur'an 2:213). The Creator and His creation must be at the central focus of Tawhid.

Finally, the author has been able to show successfully that apart from the religious rules (ahkam) that are common to the Qur'an and shari'ah, much of the worldly implications of Tawhidi methodological worldview that arise from the ontology and episteme of unity of knowledge are unknown to most of the shari'ah scholars and shari'ah protagonists for a long time. This book may act as a watershed.

M. Shamsher Ali

Professor Emeritus, Southeast University, Past President of Bangladesh Academy of Sciences (BAS), Founder Vice Chancellor, Bangladesh Open University, Founder Vice Chancellor, Southeast University

Preface

Introduction

The shari'ah scholars and their schools (sects = madhabs) and human misinterpretation of Tawhid as the LAW have a disabling mental construct and worldview that are always expected to arise from the Qur'an and the sunnah and the intellectual and applied discourse by the learned and all participants. Shari'ah scholars, sects (madhabs), and their protégé in present times understand by Tawhid just its cast of belief that is impersonal in nature, not having any social implications except for the hidden individuation of Tawhid in self. Hence, Tawhid as the essential law of 'everything' in its conceptual and applied implications has never occurred to the Muslim mind over all times. Even when Ghazali referred to the emanation of the purpose and objective of the shari'ah, this was in terms of his explanation of the dynamics of Tawhid as consciousness (taqwa and tawakkil) (Karim, n.d.). But this primordial ontological premise of Islamic monotheistic worldview in the scheme and order of 'everything' was never invoked to explain the unfathomed capability. That is to interrelate the organic monotheistic unity of knowledge with the generality and particularities of the world-system along with all the ensuing details. Muslim scholars along with their sects and humanly concocted interpretations of worldly events in the light of Tawhid have failed grossly in carrying forth the Tawhidi origin of maqasid al-shari'ah thought as it appealed to Ghazali. Yet this appeal was never formalized into conceptual and applied perspective by latter days' Muslims. The clamour around shari'ah in respect of the worldly matters (muamalat) now remains an empty secular pursuit of mind and matter of a disengaged Muslim society—not a truly Qur'anic one. The present work is thereby a serious critique of the school of shari'ah in the face of the indelible Tawhid as law in respect of its methodological worldview of unity of knowledge (monotheism) as the primal ontology of 'everything' that thereby emanates.

Reflecting on the Field of Islamic Economics and Finance as It Has Evolved

In the field of Islamic economics and finance and social ordering, it has now been 19 years since the birth of the unique educational institution of postgraduate programme in Islamic Economics and Finance (IEF) now with the Faculty of Economics of Trisakti University, Jakarta, Indonesia. This institution (see online https://www.ief-trisakti.ac.id/) was established by its academic founders, late Professor Dr Sofyan Syafri Harahap, Department of Accounting, Faculty of Economics, Trisakti University, and Professor Dr Masudul Alam Choudhury, author of this book. The specific objective behind this great educational project was to derive, develop, and apply the methodology of Tawhid (monotheism as law) as in the Qur'an and the sunnah (teachings of Prophet Muhammad) towards establishing the methodological foundation of Islamic Economics and Finance. Such an erudite venture was absent among Muslims for a long time now.

This book is a manifestation of the depth and highly analytical nature of the study of Tawhid as the primal ontological law of monotheism that is followed in IEF. It takes a deeply Qur'anic exegesis, philosophical, and specialized socio-scientific inquiry in deriving, developing, and empirically applying the imminent Qur'anic methodology of unity of knowledge and its consequential unified world-system in extensive diversity, by generality and particulars.

The approach of the study, while being of Qur'anic ontological origin, is also of a comparative nature based on critical thinking. Educational faculty, advanced scholarly students, and the global academic forum are being educated substantively in Tawhidi methodological worldview. The emergent approach is formalized in and by the modelling of the theory that is referred to in this book as Tawhidi String Relation (TSR).

The comprehensive moral and material, with abstraction and empirical inclusiveness of the nature of Tawhidi methodology of the worldview of

Qur'anic monotheism and the unified world-system by way of organic unity of knowledge, is explained by the three substantive parts. These are the Qur'anic exegesis on the study of ontology as the philosophical socio-scientific origin of Being and Becoming of Reality. Next follows the inquiry of the Qur'anic ontological foundation by the epistemology of derivation of knowledge and its nature and concrescence. Ontology and epistemology are followed by the study of phenomenology as the epistemic integration in reference to Qur'anic exegesis of emergent ideas in their generality and particulars. This integrative inquiry results in the pertinence of appropriate methods of analysis. Methodology thus leads into method of advanced analytical study and empirical application in a vast area of socio-scientific inquiry. TSR is thus rendered as the methodology combined with methods of advanced study of the theory of everything in the socio-scientific domain.

The Ontological Context of Tawhid as Law

This work along with many other ones by the author, while being derived from the ontological foundation of the Qur'anic worldview further explained by the teachings of the Prophet Muhammad (sunnah) and discussed by the learned ones in the Qur'an and sunnah (ululalbab), is a subtle extension of the meaning of Tawhid as a methodological worldview to 'everything'. The concept of 'everything' bears the meaning of universality and uniqueness of the generality and particulars of the world-system. The Qur'an refers to such a holistic domain as a'lameen. Its relational characteristics of systems, cybernetics, and analytical dynamics in the framework of complementary relationship between variables, all of the endogenous type, are described in the profundity of organic interrelations. The Qur'an refers to this property of the consciously evolutionary learning nature of the world-system in all its manifestations as both—organic interrelations between all things as created by divine ordainment (azwajakullaha), and the reoriginative phenomenon of evolutionary learning into the ever new and discoverable reality (khalq in-jadid).

The meaning of Tawhid is thereby much beyond, and is subtle and richer than the mere traditional understanding, that is, of Tawhid al-Rububiyya (creatorship by Allah's sole Lordship as absolute and all-knowing) and of Tawhid al-Uluhiyyah (Allah being the sole owner of His attributes referred to as Sifat, and the beatific names referred to as Asma). The conveying of Tawhid in the framework of its methodological worldview treats all of the above divine properties as perfection that resides solely with Allah alone. They can only be invoked but cannot be comprehended in the worldly construction of the theory of evolutionary learning in unity of knowledge (divine oneness of knowledge). The principle of oneness as unity of knowledge and its worshipping form in terms of the unity of the world-system indeed are derived from the primal ontological basis of the Qur'an. The Qur'an (6:101) refers to this foundation of the theory of 'everything' as 'primal origin': "To Him is due the primal origin of the heavens and the earth: How can He have a son when He hath no consort? He created all things, and He hath full knowledge of all things."

From the above discussion we conclude that the inalienable perfection of Tawhid cannot be used for any form of worldly being, for Allah has no exactness and similarity, no variability, and no corporeality. These otherwise are signs of deficiency, whereas Allah is not deficient. He is All-Knowing and All-Hearing. Therefore, the primal origin of Tawhid in the Qur'an, and transmitted by the Prophet Muhammad through his teachings and traditions, and further on explained by discursive function of the learned ones (ululamr), can only be used as ontological beginning to establish the nature of methodology of the Tawhidi worldview by unity of knowledge. Indeed, as pointed out by Morrow and Vittor (2009, p. 48): "Although Tawhid is typically translated as 'oneness', the term derives from the Arabic root wahada which means 'to unite, to join, to combine, and to gather.' Although Allah is absolutely and utterly One, and Islam adheres to the strictest and purest form of monotheism possible, Tawhid literally means 'unity' which means 'to bring together'."

The Tawhidi methodological worldview and thereby TSR that emanates in 'everything' is a theory of Tawhidi unity of knowledge and its embedding in the order and scheme of the world-system, known and unknown with potentiality. TSR is a methodological way of explaining the derivation and functioning of the theory based on divine Oneness as explained by the derived theory of unity of knowledge in terms of abstraction, consciousness, and application. The Qur'an explains all these functions of knowledge in socio-scientific inquiry and discovery in the midst of organic relations. Such complementarities explain inter-variables relationship in unity of knowledge. It is a property that prevails in the formalism of both knowledge (truth) and 'de-knowledge' (falsehood by its own system of relations, ikhwatul-shaiatin).

The Methodology Derived from the Theory of Tawhidi String Relations (TSR)

The study of TSR now commences as a theory of Tawhidi precept of unity of knowledge, explaining its opposite complementation in 'de-knowledge', and thereby presenting the nature of the unifying world-system in its concept of 'everything'. Truth and falsehood as knowledge and de-knowledge, respectively, exist as opposite realities in 'everything' in respect of organic unity or as dialectical dualism and methodological individualism, respectively.

The specific socio-scientific disciplines, such as economics, finance, science, and society, are taken up for a uniform study within the transdisciplinary order and scheme of 'everything' as erected by Tawhidi methodological worldview. This approach is that of the theory of unity of knowledge (truth and falsehood disparately distinct). It remains universal and uniquely embedded in explaining 'everything'. The emergent theory of the specific disciplines is thereby investigated uniquely, but with diversity of issues and problems, with the use of the ever-advancing supremely analytical majesty of rigour, richness, and critical realism. The emergent methodology thereby unifies the ontological, epistemological, phenomenological, and sustainability methods conformable strictly to Tawhidi methodology of unity of knowledge.

These methods and the formalism arising from Tawhidi methodology are thereby complex and analytical in nature. Thereby, analytical formalism becomes inevitably the structural nature of this most original work on Tawhid and the abstract and applied perspectives of the emanating world-system. Shari'ah and its components, fiqh and fatawa, are not foundational and therefore not methodological origins of Tawhidi worldview. Shari'ah and discourse on it make sense if they originate in Tawhid as the primal ontological law. It then takes its forms of accepted choices and avoidance of false choices in respect of Tawhid as the law with its foundational methodological worldview. This indeed is the historical consciousness (Lucaks, 1968) that this work intends to present in a highly analytical way.[1]

[1] Qur'an (16:36): For We assuredly sent amongst every People a messenger, (with the Command), "Serve Allah, and eschew Evil": of the People were some whom Allah guided, and some on whom error became inevitably (established). So travel through the earth, and see what was the end of those who denied (the Truth).

There is no intellectual discord between Tawhid as law and shari'ah as sub-law in as far as there are common grounds of identified Qur'anic choices arising from the former. In every other case, the choices in maqasid al-shari'ah that are not derived in the most extensive sense of the overarching law of Tawhid and do not comply with such monotheistic unity of knowledge established by and in the Qur'an remain debatable ones. They do not form a part of the Qur'anic law and can be subject to rejection and dispute. This is what has happened with the vagaries regarding several juristic edicts of shari'ah.

Reflection on the Problem of Maqasid Al-Shari'ah and Outlook on Tawhidi Methodology

Maqasid al-shari'ah remains incomplete in its comprehension of the total reality by a Qur'anic emergent methodological worldview of unity of knowledge and its induction of the world-system in space and time. For instance, as pointed out and examined in this work, the five elements of the traditional meaning of maqasid al-shari'ah remain incongruent in the light of Tawhidi methodological worldview of the generality and details of the universal world-system. This deficiency in maqasid al-shari'ah arises from its failure to develop the interactive, integrative, and evolutionary learning process orientation of the conscious universe signified between the n-number of expanding and comprehensive elements of total reality. Thus, the constriction of the knowledge and knowledge-induced space and time universe in maqasid al-shari'ah is annulled by the sustainable continuity of Tawhid as law. This comprehends the abstraction and application of Tawhid as the law of unity of knowledge between diversity of the good things while avoiding the forbidden ones.

The ontological principle of historical consciousness further emanates from the Qur'anic verse (38:27): "Not without purpose did We create heaven and earth and all between! that were the thought of Unbelievers! but woe to the Unbelievers because of the Fire (of Hell)!"

The philosophy of history along which the recurrence of continuously regenerating events describes the conscious historical path is distinct from the explanation given by Maritain (1973) and all of western interpretation of philosophy of history in that there is no interrelationship in these between divine law and worldly actions. Morality is understood as a secularized human interpretation that does not arise from the divinely ordained law. See, Maritain, J. (1973). *On the Philosophy of History*, ed. (J. W. Evans). Augustus M. Kelley Publishers, Clifton, New Jersey.

Thereby, the unique methodological worldview applies as well to 'de-knowledge'-induced dichotomy of the contrasting dualistic world-system and all that this has. The emergent path of the most original and contributory work in the world of learning presented in this work is thereby of a seriously rigorous and analytical socio-scientific nature of the theory of 'everything'. It is governed by the Tawhidi methodological worldview by its model of Tawhidi String Relation (TSR).

This work proves by logical formalism of mathematics that the only true reality is a distinctive division of a point and its monotonic transformation in the total space of Truth and Falsehood. A number of mutually separable cases exist for the coordinate position of the point, say, 'a' in the total reality super-space denoted, say, by Ω as an open universal set (mathematical super-topology). The monotonicity of 'a' is defined by the continuous performance of the function $S(a)$ and its transformations in Ω. We need to determine the position of the point 'a' in Ω with the continuous functioning of $S(a)$. Several such cases arise with their possibility and impossibility.

If 'a' belongs continuously to both truth set 'T' and falsehood set 'F', then $T = F$ everywhere. Every monotonic functional transformation by $S(a)$ in this case maintains the identity mapping of $T = F$. This implies the result that 'T' both prevails in itself as truth-choices and identifies the truth of the prevalence of 'F'. Thus, truth is unique discarding falsehood in Ω with continuity of $S(a)$.

If 'a' belongs to the mathematical union of T and F realities of functioning in Ω, then 'a' belongs to the non-void mathematical union. Now there are two possibilities. The common point 'a' of T and F in Ω and its continuous monotonous transformation by $S(a)$ implies undecidability, which is annulled by the continuity of evolutionary learning towards 'T' and annulment of 'F', as the essential nature of truth in Ω by the mapping $S(a)$ in Ω.

If 'a' belongs to the intersection of 'T' and 'F', then two possibilities arise in Ω by the mapping of $S(a)$. The common point between 'T' and 'F' can be a singleton. The common point can belong to the proper intersection of 'T' and 'F'. In both of these cases, evolutionary learning in unity of knowledge in Ω with $S(a)$ causes the 'a' to belong to the null set between 'T' and 'F', thus proving the permanent disjointedness of these sets by evolutionary learning in unity of knowledge of Ω and $S(a)$ (i.e. (Ω, S)). The prevalence of 'T' over 'F' in Ω, that is, functionally, $T \cup F \Rightarrow T(F) = T$. Every monotonic '$T$' function of $T \cup F$ is 'T'. The above formalism establishes the uniqueness of the model of unity of knowledge that is derived from the Tawhidi monotheistic ontology and conceptualized and applied to both the evolutionary knowledge super-space and the 'de-knowledge' world-system, though with the differences in the properties of these two opposite worldviews in *T* and *F*.

This book is an extensive analytic study of the three cases relating to '*a*' in (Ω, S). The occurrences in (Ω, S) as Tawhidi law differentiating '*T*' and '*F*' in respect of the unique truth statement of '*T*' as the totality of Qur'anic worldview that governs all aspects of reality as truth occurrences and logical possibility, yet not so the elements of falsehood choices and actions. These latter entities are differentiated between '*T*' and '*F*' in (Ω, S).[2] This work derives the space of falsehood as 'de-knowledge'. It is marked by the singular character of differentiation and competing trade-off. Both of these characters of 'de-knowledge' mark the attributes of marginal substitution, socio-scientific Darwinism, and methodological individualism that are all contrary to the precept of Tawhidi unity of knowledge, moral consciousness, and complementarities (participation) in the good things of life while avoiding the false choices and acts.

This work and its erudition resemble the content of the book entitled *Relativity, the Special and the General Theory* by Albert Einstein (trans. R. Lawson, 1954). Yet by its originality of derivation strictly from the Qur'an, the present work is of a distinct and original nature. It confronts as the subtle façade in contradistinction to the rationalist mind of space–time concept. This reasoning is thoroughly replaced by Qur'anic logical formalism of knowledge, space, and time dimensions induced by unity of knowledge.

In the same way, the primacy of Tawhid as law is formalized analytically in this book to replace the narrow and constricted reasoning, erudition, and domain of shari'ah approach to the study of socio-scientific phenomenology. In the uncompromising sense of universality and uniqueness of Tawhid as law of 'everything' with its distinctive fullness, the Tawhidi methodology represented as theory of TSR replaces both occidental rationalism and shari'ah approach to nothing in the Tawhidi theory of 'everything'. The phenomenological understanding of methodology combines the primal ontological worldview of unity of knowledge with the imminent epistemology to construct the functional ontology and episteme in

[2] Qur'an (41:42): "No falsehood can approach it from before or behind it: It is sent down by One Full of Wisdom, Worthy of all Praise."

the ever advancing evolutionary learning processes. Thereby, methodology is the formalism underlying continuous nexus of normative and positivistic worldviews into an organically unified one. Reasoning and analysis thereby emanate from the embedded premise of philosophy of science (Blaug, 1993; Fox, 1997; Pheby, 1988). Such origination of reasoning and structure in logical formalism leads into the phenomenology of modulation and analysis.

The Abstracto-Empirical Dimension of Tawhidi Methodology

The consequential holism here is referred to as abstracto-empirical dimension of Tawhidi methodological worldview. Such a unified methodological worldview that is central to the Qur'anic organically unified universe and its details was nonetheless not practiced by Islamic thinkers by and large (quotes follow from Nusseibeh (2017)). For instance, Ibn Taiymiyyah pronounced the baseless remark, "that Islam has no philosophers". On the other hand, Not only did Avicenna free himself from preset systems and styles, he also pursued reason until he arrived at its limits. The wholly peripatetic approach was also the exclusively speculative devotional methodology of Ghazali. To the present day the singularly important theme of methodology that would apply universally to a theory of 'everything' in the individual (household) and socio-scientific domain has remained foreign to Muslim scholars (Al-Faruqi, 1982a, 1982b; Al-Attas narrated by Daud, 1998; Bakar, 1991, 2014; Nasr, 1978). All the contemporary ink of Islamicization and Gnosticism could not yield Tawhidi, and thereby Qur'anic methodology of any substantive shape, form, and schema. Consequently, the truly socio-scientific advancement receded into oblivion. Thus, the failure of Islamicization of knowledge was pointed out by Rahman (1988).

This book takes the viewpoint of logical formalism that, if the shari'ah was to be enforced as the Islamic law, then the question is, what is the status of Tawhid as the ontological law of 'everything'? It is illogical that both of these premises would compete as the law and that the shari'ah manifests Tawhid as law. The universe and the Prophet Muhammad were sent by one law that ideally represents the totality of the Qur'an, transmitted in interpretation by the sunnah and the discourse of the learned (ululamr) based strictly on the Qur'an and the sunnah—not on madhabs and humanly interpreted reasoning in the first place.[3] It is regrettable that today neither do Muslims by and large invoke learning by questioning concerning Tawhid and the world-system nor do the Muslim clergies devote to the study and engagement of the Islamic knowledge and practices on the penetrating depth of Tawhid and the world-system for exciting Muslim erudition.

This work has advanced a step further than the earlier ones by this author on the role of knowledge induction in the attenuating multivariate system and the resulting empirical interpretation. This work has deconstructed knowledge into its interior component of belief. Thus a good deal of analytical sections deals with the functioning of the belief-induced (ε) knowledge parameter in the form, $\{\theta(\varepsilon)\}$. Despite this, belief is treated as an inner dynamic force, not as an empirical effect. Conceptual implications of belief-centred knowledge induction predominate.[4]

Jakarta, Indonesia — Masudul Alam Choudhury

References

Al-Faruqi, I. R. (1982a). *Tawhid: Its Implications for Thought and Life*. Herndon, VA: International Institute of Islamic Thought.

Al-Faruqi, I. R. (1982b). *Islamization of Knowledge: General Principles and Workplan*. Herndon, VA: International Institute of Islamic Thought.

Bakar, O. (1991). *Tawhid and Science, Essays on the History and Philosophy of Islamic Science*. Kuala Lumpur, Malaysia: Secretariat for Islamic Philosophy and Science.

Bakar, O. (2014). *Islamic Civilisation and the Modern World*. Darassalam, Brunei: University of Brunei Darassalam Press.

Blaug, M. (1993). *The Methodology of Economics*. Cambridge: Cambridge University Press.

[3] Qur'an (69: 44–47): "And if he (Muhammad) had forged a false saying concerning Us (Allah), We surely should have seized him by his right hand (or with power and might), And then certainly should have cut off his life artery, and none of you could withhold Us from (punishing) him."

[4] Qur'an (41:14): The desert Arabs say, "We believe." Say, "Ye have no faith; but ye (only) say, 'We have submitted our wills to Allah,' For not yet has Faith entered your hearts. But if ye obey Allah and His Messenger, He will not belittle aught of your deeds: for Allah is Oft-Forgiving, Most Merciful."

Daud, W. M. N. W. (1998). Metaphysical Worldview. In *The Educational Philosophy and Practice of Syed Muhammad Naquib Al-Attas, an Exposition of the Original Concept of Islamization* (pp. 33–68). Kuala Lumpur, Malaysia: International Institute of Islamic Thought and Civilization.

Einstein, A. (1954). The Problem of Space, Ether, and the Field in Physics. In C. Commins & R. N. Linscott (Eds.), *Man and the Universe: The Philosophers of Science*. New York: The Pocket Books, Inc..

Fox, G. (1997). *Reason and Reality in the Methodologies of Economics*. Cheltenham: Edward Elgar.

Karim, M. F. (Trans.). (n.d.). Imam Ghazali's *Ihya Ulum-Id-Din in 5 vols* (esp. Vol. 4). Lahore: Sh. Muhammad Ashraf.

Lucaks, J. (1968). *Historical Consciousness*. New York, NY: Harper & Row Publishers.

Maritain, J. (1973). *On the Philosophy of History* (J. W. Evans, Ed.). Clifton, NJ: Augustus M. Kelley Publishers.

Morrow, J. A., & Vittor, L. A. (2009). Tawhid in Theological Mode. In Al-Shaykh al-Saduk, J. A. Morrow, A. R. Rizvi, L. A. Vittor, & B. Castleton (Eds.), *Kitab al-Tawhid, The Savor Foundation* (pp. 35–49). Paradstesh.

Nasr, S. H. (1978). *An Introduction to Islamic Cosmological Doctrines*. Boulder, CO: Shambhala.

Nusseibeh, S. (2017). The Nature of Truth. In *The Theory of Reason in Islam* (pp. 167–168). Stanford, CA: Stanford University Press.

Pheby, J. (1988). *Methodology and Economics, a Critical Introduction*. London: Macmillan.

Rahman, R. (1988). Islamization of Knowledge: A Response. *American Journal of Islamic Social Sciences, 5*, 1.

Contents

List of Figures

CHAPTER 1

Introduction

Qur'anic Elan on Tawhid (Monotheism) and Shari'ah (Shari'atan)

Introduction

This book by invoking the Tawhidi ontological foundation of the Qur'anic law and worldview is also a study on ta'wil, meaning deepening in exploring the hidden meaning of Qur'anic verses. On this aspect of learning the Qur'an, there is the hadith of the Prophet Muhammad as follows[1]: "The Quran possesses an external appearance and a hidden depth, an exoteric meaning and an esoteric meaning. This esoteric meaning in turn conceals an esoteric meaning so it goes on for seven esoteric meanings (seven depths of hidden depth)." A saying of the Prophet Muhammad goes as follows: "In truth, the Qur'an possesses an inner and outer, and its inner possesses another inner up to seven inners."[2] Furthermore, there is a hadith of Prophet Muhammad that states the Qur'an has an inner meaning, and that this inner meaning conceals a yet deeper inner meaning, and so on (up to seven successive levels of deeper meanings), has sometimes been used in support of this view.[3]

It is only ta'wil of the overarching meaning of Tawhid as law in the Qur'an that the socio-scientific study of the Qur'an covers the universal

[1] H. Corbin (1993). *History of Islamic Philosophy*. pp 1–14. Kegan Paul International, London, England.

[2] Sayyed haydar 'amoli, Jami-u al-asrar (Shirkatintisharat 'ilmifarhangi, Tehran, 1989), p. 530. The hadith narrated from the Prophet.

[3] Wikipedia, "Esoteric interpretation of the Qur'an" (visited March, 2019).

M. A. Choudhury, *Tawhid and Shari'ah*,
https://doi.org/10.1007/978-3-030-49087-4_1

meaning,[4] its explanation, and application of Tawhid as law as the way and substance that is implied in the context of the term 'shariatan' in the Qur'an. When the meaning of the term 'shari'ah' denotes the common Qur'anic law and the universal and unique worldview of Tawhid as law, then the meanings of Tawhid and shari'ah coincide. Other than this common worldview of Qur'anic law, shari'ah over the ages became a sectarian and human-concocted law of jurisprudence. In this form, neither the resulting extraneous nature of juristic interpretation by maqasid al-shari'ah nor shari'ah-compliance are acceptable within the Qur'anic fold. The unique focal point is that the Qur'anic meaning of Tawhid and 'shariatan' must be unequivocally premised on the worldview of unity of knowledge that the Qur'an has explicated in terms of organic and symbiotic embedded relationships between all things in pairs (pairing of the universe). Without this unique foundation of knowledge and worldview it is impossible to learn from the Signs of Allah by unity of knowledge in the order and scheme of 'everything'. The human-concocted origin of shari'ah as jurisprudence and its rampant practice today is permanently oblivious and unknowing about this cardinal ta'wil of the Qur'anic worldview of Tawhid in terms of its meaning of symbiotic pairing of the universe in its generality and details (Qur'an 36:36).

The equivalence of shari'ah and Tawhid is therefore stated as follows:

Tawhid = Shari'ah = Way of arising from learning in and leading towards the understanding of monotheistic oneness in the scheme and order of the symbiotic universal reality.

Hence, the emergence of Qur'anic socio-scientific thought is ontologically premised in the ta'wil of Qur'anic verses. Thus, Tawhid as the pairing of the generality and particulars of the universe is the ta'wil meaning of organismic and symbiotic relations of oneness. 'Shariatan' implies the way derived from and towards confirmation of Tawhid as universal law 'between the heavens and the earth and all that is below it and above it'. Such depth of learning and discovery is by ta'wil the ways of understanding the signs of Allah in the order and order and scheme of 'everything'

[4] The term 'ta'wil' is invoked by the Qur'an (12:100) to mean such lofty meanings of the inner understanding of verses in the light of the most fundamental truth. That is, Tawhid as the universal law of 'everything'.

(Qur'an 41:53)[5]. Shari'ah as jurisdiction fails in this by its absence of premising knowledge explicitly in Tawhid as the paired unity of knowledge extending between the heavens and the earth by symbiotic manifestation.

Indeed, Tawhid is ontologically primal. It dawned as reality in the universal order at the moment of creation (kun fa-yaqun in the Qur'an). At the simultaneous moment of creation, Tawhid as total law depicting the signs and understanding of the oneness of Allah in the order and scheme of things was revealed in two of its integral forms. Firstly, it entered the domain of intrinsic belief in monotheism (iman in the Qur'an). Simultaneously, it expatiated as organic unity of knowledge in the generality and particularity of the world-system in respect of the signs of Allah. The two elements together form the universal and unique functioning of Tawhid as law in 'everything'. Now if we denote $\{\theta\}$ as the levels of comprehending unity of knowledge in respect of the signs of Allah in everything, and $\{\varepsilon\}$ as the degree of belief, then the embedding of belief in knowledge causing the configuration of knowledge and the world-system can be denoted by $\{\theta(\varepsilon)\}$. Since $\{\varepsilon\}$ remains intrinsic but not measured and thus not parametric, so Tawhid as law is functionally configured for abstracto-empirical work by measured consciousness at the level of $\{\theta\}$. $\{\varepsilon\}$ remains implicit and unmeasured quantitatively at the level of moral construction of the social order. Yet it is of fundamental importance in the invoking of the true power of ta'wil in understanding Tawhid in the order and scheme of 'everything'. We thereby write in recursively continuous and evolutionary meaning of unity of knowledge and the world-system:

$(\Omega, S) \leftrightarrow \{\varepsilon\} \leftrightarrow \{\theta(\varepsilon)\} \leftrightarrow \{x(\theta(\varepsilon))\} \leftrightarrow$ continuity by the endless exercise of ta'wil of the Qur'an and the sunnah (Ω, S) for manifestation by law in the world-system. $\{x(\theta(\varepsilon))\}$ denotes the world-system in generality and details induced by unity of knowledge. Recursive continuity as sustainability of Tawhid in knowledge, space, and time continuum is denoted by pairing, $\leftrightarrow$.

If the compound mapping over the entire string shown above and denoted by $\leftrightarrow$ is denoted by F, then ta'wil in the mathematical sense is explained by,

[5](Qur'an 41:53): We will show them Our signs in the horizons and within themselves until it becomes clear to them that it is the truth. But is it not sufficient concerning your Lord that He is, over all things, a Witness?

$$dF(\theta(\varepsilon))/d\theta = \alpha(\theta(\varepsilon)).F(\theta(\varepsilon));$$
$$\text{or } F(\theta(\varepsilon)) = \int_{\theta(\varepsilon)} A(\theta(\varepsilon)).F(\theta(\varepsilon))^{\alpha(\theta(\varepsilon))}\, d(\theta(\varepsilon))$$

This formal mathematical expression of ta'wil means that ta'wil is ontologically premised in the precept of Tawhidi unity of knowledge as conveyed by the learning string in terms of $F(\theta(\varepsilon))$. Ta'wil is the learning experience in Tawhidi unity of knowledge in deepening depth of its meaning and application. It is so heightened across the knowledge, space, time, dimensions that are all embedded in unity of knowledge $\{\theta(\varepsilon)\}$. The recursive symbol ↔ establishes the case of equivalence and circular causality along the entire evolutionary learning path, the TSR, in respect of Tawhid and the world-system in continuum. $A(\theta(\varepsilon))$ and $\alpha(\theta(\varepsilon))$ in the above expressions denote positive coefficients of the force of learning in terms of Tawhidi unity of knowledge along the belief-induced evolutionary knowledge induction caused by continuous exercise of ta'wil.

Neither maqasid al-shari'ah nor the shari'ah-compliance idea has been able to launch an analytical study of the imminent Qur'anic socio-scientific world-system extending between the 'heavens and the earth, and across all that is below the earth and above the earth'. The present work unravels such a methodology by way of using the approach of ta'wil of the Qur'anic centrepiece of Tawhid and its derived methodology as a theory of the universal and unique law in the Tawhidi ontological framework of organic unity of knowledge and the world-system. The contents of this work are divided between theory and applications to current topics. A comparative and critical study of Tawhid as Qur'anic law contra shari'ah as Islamic jurisprudence is maintained in the formalism of theory and application.

The language of the text is rigorously analytical spanning areas of ta'wil of the Qur'an and sayings of the Prophet Muhammad (hadith). This includes analytical philosophy of science, specialized knowledge of the world-system of Islamic economics and society, quantitative analysis and implications, comparative study in these areas between Islamic and contemporary scholarship. Such a work of bold and rigorous erudition, though not in compliance with the outmoded Islamic traditional understanding and the school of theocracy, is in full consonance with the Qur'an and sunnah. The book is critical of the sectarian (madhab) conception of relational independence of facts. Thus, the non-creative outlook of shari'ah contrasts with universality and uniqueness of Tawhid as the analytically established law explaining the monotheistic organic unity of being and becoming in 'everything'. This wide and strict methodological development of the Tawhidi worldview is articulated in this work. The book is thereby an erudite contribution to the world of learning that would interest all in scholarship who would like to understand the ontological analytical foundation of ta'wil in the Qur'an concerning the imminent socio-scientific universe of unity of knowledge as theory and application.

CHAPTER 2

A Critique of Shari'ah Perspectives Contra Tawhid as Law: Terminologies

The Structure of Tawhid as Law of Unity of Knowledge and the Embedded World-System

The Tawhidi methodological worldview has its most distinctive characteristics that define the all-comprehensive domain of 'everything' and thereby configure the revolutionary socio-scientific shape of things in their normative, positive, and the future possibilities. These distinctive features of the nature and logic of Tawhid as the primal ontological law (Qur'an, 2:117)[1] and its moral and material abstracto-empirical extant of knowledge are as follows:

Firstly, the expression of the strictly monotheistic oneness as inherently manifest in the scheme and order of 'everything' is explained by the principle of unity of knowledge between the good things of life as truth, while avoiding the dejected things of life as falsehood. This attribute is sustained in continuums of moral-material order of socio-scientific reality in abstracto-empirical forms.

Secondly, the formalism derived from the ontological foundation of Tawhid as unity of knowledge induces the generality and details of 'everything' in the world-system. This is manifested by the principle of universal complementarities (systemic participation), separately between the good choices and the dejected choices in their own distinct experiences.

[1] Qur'an (2:117): "To Him is due the primal origin of the heavens and the earth: How can He have a son when He hath no consort? He created all things, and He hath full knowledge of all things."

M. A. Choudhury, *Tawhid and Shari'ah*,
https://doi.org/10.1007/978-3-030-49087-4_2

Thirdly, the concept of unity of knowledge as the manifestation of the nature and logic of monotheistic unity of being and becoming is represented as ontological worldview and phenomenological (consciousness) material and experiential application by organic interrelations of participative complementarities by positive knowledge induction. Thus, the variables of vectors defining 'everything' are characterized by their dynamic interrelationships marked by continuous organic inter-causality between the inter-variables of vectors. The same feature is true of the false things of life, except by virtue of the pervasive nature of their inter-variable differentiating dialectics called bifurcation of the world-system (Wallerstein, 1998). Thereby, the law, formalism, and applications of Tawhid as law in respect of its moral and material expressions in the world-system assume universality and uniqueness of explanation in 'everything'.

Fourthly, spanning of events in the generality and specifics of the complex order of interrelated world-systems of 'everything' is analytically explainable within and across the knowledge, and knowledge-induced space and time dimensions. This extant of evolutionary knowledge generation sustains over eternity—from the Beginning to the End by their recursive equivalence in terms of the supercardinal nature of absolute and complete knowledge of the monotheistic stock of complete knowledge as primal ontological phenomenon. For the case of the expanse of 'de-knowledge', the entropy of this system is self-destructive and, thereby, finitely contained. These are properties of knowledge and 'de-knowledge' derived from the Qur'an (7:191–192).[2]

Fifthly, the entire evolutionary knowledge-induced conscious historistic path from the Beginning to End in terms of the equivalence of Closure by the Great Event (Qur'an)[3] is defined, formalized, explained, and applied by learning processes that perpetually differentiates between the good choice and the bad ones. These evolutionary learning processes assume their descriptions within and across bundles of intra-system and inter-system temporality with complexity of systemic inter-variable causality leading to the evaluation of inter-variable discursively determined objective criterion of wellbeing (maslaha). Every such event is explained by evolutionary learning processes that are characterized by interactions, integrations, and evolutionary (IIE) learning across the knowledge, space, and time dimensions.

[2] Qur'an (7:191–192): "Will they associate [with Me] those that do not create, but are [themselves] created—those that can bring them no victory?"

[3] Qur'an (78:1–5) mention of the Great Event of the Qur'an, Tawhid, End, Prophet Muhammad (Risalah).

In summary to the above points, the principal attributes of Tawhidi law are its methodological characterization of a knowledge-induced world-system. The resulting attribute of Tawhidi law in respect of the monotheistic worldview of unity of knowledge is a functional design concerning Tawhidi law as a moral-material embedded use of a worldly intellection. In this respect Tawhid as law is an abstracto-empirical organization of phenomenological (consciousness) thought. The consequential outcome of the resulting generality and details of the attenuating study of the multi-stage world-system by way of the wellbeing function, subject to its circular causal relations of the ascribing variables, conveys the meaning of non-optimal, complex, and evolutionary learning process. The continuous IIE properties are invoked as conveying new outlook to analytical socio-scientific study. The emergent methodological approach characterizes correctly the transdisciplinary nature of Islamic economics, finance, and process-based socio-scientific study. Such an approach which is a logical analytical outcome from the Tawhidi primal ontological foundation of unity of knowledge and its formalism is contrary to the optimization calculus in conventional socio-scientific analytics. Shari'ah and its proponents in shari'ah-compliance economics, finance, and socio-scientific studies have subjected themselves to the conventional methodology that is contrary to Tawhidi methodological worldview.

The Contrary Worldview of Shari'ah

The equivalence between Tawhid as the ultimate law invoking the Qur'an and sunnah and the contending belief in maqasid al-shari'ah as law is only in terms of those rules (ahkam) that are common to both of these fields in reference to the Qur'an and sunnah and their exegeses. Other than this common domain of knowledge any aspect of the shari'ah remains to be subservient to shari'ah as human-concocted juristic developments called fiqh and the legal preferences based on fiqh called usul al-fiqh. In these matters, many of the ahkam on product developments and exegetic interpretations are found in the history of shari'ah to have emerged from Muslim sects called madhabs, and conveniences of satisfying product development by discourse between government-appointed shari'ah bodies and the lobby groups. In these groups were the rulers, elitists, and preferred shareholders of Islamic wealth and power. Such incidents were found in the courts of the Fatimide and Umayyad rulers, and in all other kings and governing madhabs. Such a state of shari'ah rule was noted to the disdain of Ibn Khaldun (Mahdi, 1964). Imam Ghazali found it to be an untrustworthy practice of the public authority during his time. Here is what the Stanford Encyclopedia of Philosophy (Ghazali, visited Oct. 2018) writes about Ghazali's disposition to the courts and the misused regal power of state machinery: "He realized that the high ethical standards of a virtuous religious life are not compatible with being in the service of sultans, viziers, and caliphs. Benefiting from the riches of the military and political elite implies complicity in their corrupt and oppressive rule and will jeopardize one's prospect of redemption in the afterlife. When al-Ghazâlî left Baghdad in 1095 he went to Damascus and Jerusalem and vowed at the tomb of Abraham in Hebron never again to serve the political authorities or teach at state-sponsored schools."

The greatest vacuity in maqasid al-shari'ah and its subsequent derivation of shari'ah rules, as in worldly affairs of muamalat, was its continued methodological delink over time between the great cosmological matters and matters of social existence. This was not meant to be so according to the Qur'an. Contrarily, the Qur'an establishes a systemic keen organic relationship between different precincts of the cosmological order and the terrestrial order. This is true in regards to both the widest field of conceptual, analytical, and thus scholarly issues and within them. Thus shari'ah could not provide any methodological energy to develop and carry on the law of organic interrelations between the heavens and the earth, despite that the Qur'an embodies the universal and unique law to all that is between the heavens and the earth, below and above the earth. This comprehends all of 'everything' in the knowledge, space, and time dimensions (Qur'an, 13:1–4) Likewise, in terms of the allegorical but most powerful extension of the meaning of God as the Lordship to the all-comprehensiveness of the divine law (sunnat Allah = Tawhid as law), the Qur'an (11:123) declares: "And God alone comprehends the hidden reality of the heavens and the earth: for, all that exists goes back to Him [as its source]. Worship Him, then, and place thy trust in Him alone: for thy Sustainer is not unaware of what you do."

Maqasid al-shari'ah was erected on five pillars of ethicality concerning muamalat (worldly affairs) excluding the great cosmic and deeply analytical scholarship of universality and uniqueness concerning 'everything'. These pillars were endowed state powers to enact and actualize the criterion of wellbeing (maslaha) as:

1. Protection of religious knowledge. This can be assumed to be Tawhid as theological belief rather than as corporeal manifestation of the law of unity of knowledge out of organic pairing (Qur'an 36:36).
2. Protection of reason (aql). This is more accurately interpreted as the nature of knowledge (fitra).
3. Protection of life. This can be extended to the principle of sustainability at the present.
4. Protection of progeny. This can be implied to intergenerational sustainability.
5. Protection of property rights. Ownership of property rights means a trusteeship of the extended meaning of global protection of resources, real and intellectual, material and spiritual for the shared future of a just, fair, and equitable world as organic balance of entities that together define the principle and criterion of wellbeing (maslaha).

The mujtahids of the classical Islamic period gave narrative importance to the above principles in their functioning and commodity codification pertaining to muamalat alone. The latter days' shari'ah scholars understood the above principles in terms of their legitimacy as separable ones. The organic relations interconnecting them were ignored. A stark reflection of this misguided consequence can be found in respect of product development in the so-called Islamic economics and finance. The various financial instruments are their contract bases and are made to rest separately as legally accepted under shari'ah perspective. Thereby the concept of organic pooled portfolio of various financing instruments and the corresponding meaning of inter-alia complementary nature of social contracts has never taken shape and form. Rather, in some cases, shari'ah is known to obstruct jointly interdependent social contracts.

Yet the Tawhidi legal perspective must rule upon the acceptance, revision, or rejection of financial contract on the basis of the benefit of participatory balance of organic interrelationship that can be formed between diverse financial instruments and their underlying contracts. Thus, the complementary (participatory) interdependency in terms of unity of knowledge between inter-causal though diverse financial instruments and financial contracts can be legitimated. This kind of co-existing portfolio of contracts is not yet legitimated by shariah (Arbouna, 2007).

By the same kind of adverse conclusions advocated by shari'ah with respect to complementing participatory possibilities, a misguided viewpoint most probably derived from a misunderstanding or from a weak ahadith named by the shari'ah gurus, has been used to obstruct a system methodological worldview. Thereby the rich analytical depth and the endogenous participatory development of all resulting consequences that are conducive for wellbeing have been lost. The rich systemic methodological worldview of unity of knowledge arising from the Qur'an and sunnah as of Tawhid has never occurred in shari'ah and among its proponents in any of the transdisciplinary perspectives of possibilities and their functioning.[4]

This general feature of methodological independence of different goals of maqasid al-shari'ah in generating the organismic nature of the maslaha function has caused dislocation between all these goals to become isolated ones. The maslaha function could not be defined in terms of the ultimate Qur'anic and sunnah worldview of Tawhid as the primal law given its corporeal form in respect of moral-material and abstracto-applied functioning in 'everything'. In recent times as traditionally, the construction of maslaha function has suffered from exogenously enforced weights arbitrarily selected for certain preferred and selected variables. Mostly a linear form of additive maslaha function is constructed as maqasid-index. Thus, a fully juristic preference of istihsan has been conveyed to the formulation of such maqasid-indexes not having their automatic endogenous indexing arising from the nature, revision, and possibilities of the world-system with its vector of knowledge-induced variables.[5]

[4] Narrated in Sahih Bukhari hadith Book 8, Hadith 6: Narrated 'Abdullah:

> The Prophet drew a square and then drew a line in the middle of it and let it extend outside the square and then drew several small lines attached to that central line, and said, "This is the human being, and this, (the square) in his lease of life, encircles him from all sides (or has encircled him), and this (line), which is outside (the square), is his hope, and these small lines are the calamities and troubles (which may befall him), and if one misses him, an-other will snap (i.e. overtake) him, and if the other misses him, a third will snap (i.e. overtake) him." This hadith implies that Prophet Muhammad was deeply knowledgeable of the systemic nature of the methodological worldview.

[5] The endogenous automatic simulation of the maslaha function, say f_1, is explained by monotonic positive transformation by functional transforms $\{f_i\}$ wherein the maslaha function-specific evaluated (estimated and simulated) coefficients are discursively determined by appropriate forms of mathematical and statistical methods. The compound maslaha function with increasing extant of possibilities by extending variables representing goals beyond the traditional five maqasid-goals, and the statistically simulated coefficients of such variables is

In the case of the traditional maqasid-determined maslaha function, the methodological independence between the five or more goals as exogenously formulated takes the form,

$$f_i(\mathbf{x}) = \sum_{i=1}^{5} a_i \times x_i; \quad i = 1,2,3,4,5; \tag{2.1}$$

x denotes the vector of the five variables representing the five respective maqasid-goals and their extensions. $\{a_i\}$ are exogenously assigned as weights for $\{x_i\}$. In the case of compounded functions, product terms in $\{x_i\}$-variables reduce to zero, because there is no definition for complementary interdependency interconnecting the maqasid-variables. Therefore, the compounded positive monotonic function, say F(**x**), yields for two variables i = 1, 2, and likewise,

$$F(x) = \sum_{i=1}^{5} a_i \times x_i + \sum_{i=1}^{5}\sum_{j=1}^{5} a_i \times a_j \times x_i^2 \tag{2.2}$$

The implication of methodological independence between the goals (variables) could mean functional independence between (2.1) protection of religion (deen), and (2.2) protection of reason (aql, fitra). This is to be oblivious of the presence of divine knowledge (sunnat Allah) in any and 'everything'. This inference, although not implied by maqasid al-shari'ah, has nonetheless not been explicated as an article of organic endogenous relationship of (2.1) with 'everything' as a methodological worldview.

Likewise, methodological independence between, say, (4) protection of intergenerational posterity and (5) property right would mean absence of inter-causal relationship between the use of resources for posterity and in reversion the conservation of resources intergenerationally for the actualization of trusteeship and vital distribution of resources and ownership towards the wellbeing of posterity. These kinds of separated goals of maqasid al-shari'ah, while not deliberate, is the ultimate consequence of a

then endogenously determined as $[f_1 \cdot f_2 \cdot \ldots\ldots \cdot f_n](\theta)$ for i = 1, 2, ..., n, in the family of compounded goals $\{f_i\}$. The endogenous effect to such monotonic positive functions is given by the induction of knowledge 'θ' derived from the Tawhidi primal ontological law of unity of knowledge in 'everything'. An Islamic version of commodity and functioning theory for wellbeing function is thereby offered by the worldly consequences of Tawhid as law in its corporeal reality.

system that remains devoid of methodological foundation to think along systemic lines. This was the problematique of shari'ah ever since the inception of the traditional years of maqasid al-shari'ah and continues on today. Its future remains the same in the absence of a true reversion to Tawhidi primal ontology followed by epistemology and phenomenology ingrained in global consciousness and applied into 'everything' of the world-system.

The differentiating problem contra systemic organic unity of being and becoming of knowledge-induced reality as of Tawhid as law is similar to the problem of heteronomy found in Kant's episteme of reasoning (Kant, 1949). Kant as moralist and social philosopher raised the consciousness for a unison between moral imperative of a priori reasoning to be embedded with the a posteriori reasoning of the constructed world-system. Thus, the socio-scientific worldview of pervasive complementarity between moral law and material world-system remained deeply ingrained in Kant's ontology resulting into its epistemology as a purely normative idea. The part on phenomenology as the science of practical embedding between moral and material order of total reality did not result from Kant's normative desire. The normative and the positive could not therefore be endogenously synthesized in Kant. What resulted by the crevice between a priori and a posteriori reasoning on socio-scientific knowledge was the divide between the moral law and the material law, although this was not the conceptual project of Kant's deontology concerning moral and material unitary system. Consequently, morality and consciousness had to be exogenously enforced into the a posteriori realm of reasoning. Such an exogenous moral enforcement was seen as intrusion into free human will in its independent choice. This resulted in the problematique of heteronomy. It has remained permanently ingrained in all of occidental socio-scientific thought. It is the permanent nature of rationalism. The postulates of economic rationality and functional maximization are derived from the roots of rationalism. Amartya Sen called such practitioners as rational fools. Similarly, Lawson (2009) decried in his *Ontology and Economics: Tony Lawson and His Critics*, Routledge, London, UK).

The methodology of shari'ah suffers from similar problems as Kantian ontology. Shari'ah is likewise unable to unify the a priori and a posteriori domains of unified reality as an endogenously induced consequence of unity of knowledge of Tawhid as corporeal law. Consequently, the realization of moral and material complementarities as Signs of Allah (ayath Allah) of the world-system has never been possible by shari'ah as long as it continued as a secularized understanding of moral law with many sectarian variations.[6]

While expressions (2.1) and (2.2) are the results arising from the absence of the methodology of unity of knowledge, and which has found its ways into product development of Islamic finance, a reversal of maqasid al-shari'ah to the pure origin of Tawhidi law is possible. The maqasid al-shari'ah would then be transformed into the codification of commodities and functioning of the Right Way (Tawhid) that the Qur'an declares (3:51): "It is Allah Who is my Lord and your Lord; then worship Him. This is a Way that is straight."

The equivalence between Tawhid as the corporeal law of unity of knowledge explaining monotheistic oneness in terms of organic relational homogeneity (endogeneity) is explained by the following relational path of inter-causality from the Beginning of Creation to its End with everlasting sustainability of Tawhid as law across continuums:

[Creation as Beginning (Tawhid as monotheistic unity of knowledge) → Corporeal law of unity of knowledge in 'everything']
↔moral/ethical conception of a priori
↔phenomenology of knowledge-induced application (consciousness by organic unification between a priori and a posteriori)*
↔sustainability in continuums until the End (Akhira)

*stands for the case where maqasid al-shari'ah just as all of socio-scientific thought fails and can be reconstructed by the ontological emanation as stated. The final result for the Tawhidi ontological transformation of maqasid al-shari'ah is explainable in expression (2.3),

[6] The traditional shari'ah concepts based on Muslim sects (madhabs) were vehemently opposed by the reformists along lines of reforming shari'ah in the strict light of the Qur'an and sunnah. Rashed Rida (1983) ("Introduction", Al-Mughni, vol. 1, Dar al-Kitab al-Arabi, Beirut, Lebanon) wrote in this regard: "In sum, it is attested in the apodictic texts of revelation and in the consensus of the community that it is a great sin for Muslims to be divided among schools of law and in their opinions, not to mention the fanaticism of each group for its particular school whether on matters of creed or law. . .the harmful and evil consequences of such division (tafarruq) is mentioned in the history books and has led in these days to the weakening of the Muslims, the disappearance of their political rule and the domination of the foreigners over their territories." The English translation is from Haykel, B. (2014). "The political failure of Islamic law", Yale Law School Legal Scholarship Repository, Occasional Papers.

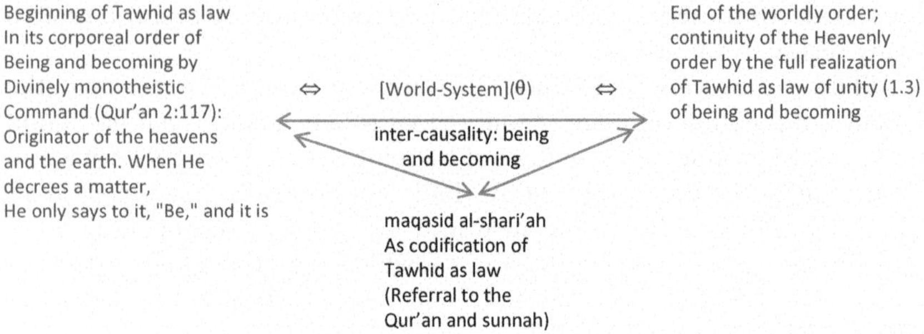

(2.3)

In reference to expression (2.3), expressions (2.1) and (2.2) are distinctly altered into the following forms:

$$f(\mathbf{x}) = \prod_{i=1}^{n} \left[x_i^{ai} \right](\theta); \quad i = 1,2,3,4,5 \tag{2.4}$$

Expression (2.4)[7] can be log-linearized with dynamic simulated coefficients $\{a_i(\theta)\}$ of respective variables corresponding to maqasid-codified goals $\{x_i(\theta)\}$ due to variations of the $\{\theta_i\}$-knowledge parameters. The GIS-Spatial Domain Analysis (SDA) method can be used as machine-generated coefficient values. A combination of discourse and SDA-generated values can be applied. Random coefficients regression methods can be applied. All of the following kinds of data can be used: time-series, surveys, and inter-meshing of these. In the case of functional compounding, the same methodical approach applies. Thereby, expression (2.2) is replaced by the log-linear form of the following functional relation:

[7] The logarithmic form of the wellbeing function conveys the pervasive principle of complementarities between the variables and their representative balance of such complementarities by the wellbeing function. For instance, the neoclassical substitution between any two variables and functions will result in, $df(\mathbf{x}(\theta))/d\theta = \sum_{\mathbf{x}(\theta)} \left(\partial f(\mathbf{x}(\theta))/\partial \mathbf{x}(\theta)\right) \times \left(d\mathbf{x}(\theta)/d\theta\right) > 0$ identically, in both cases of the f(**x**(θ))-relations as maslaha function or mafasid function ('θ' as de-knowledge). Therefore, neoclassical property of substitution in resource allocation is permanently replaced by the property of dialectical learning or 'de-learning'. Both are determined in reference to the methodological worldview of unity of knowledge or contrarily 'de-knowledge' induced dialectical processes. For instance, Sztompka conjured a dialectical model of social becoming, but denied the role of divine organistic relationship of unity of knowledge.

$F(\mathbf{x}(\theta)) = \prod_{j=1}^{m} f_j(\mathbf{x}(\theta))$. In this case there are m-number of compounded functions, each of the form (2.4) with extended variables, thereby goals, of the maqasid-codification in reference to Tawhid as the governing law. The resulting functions with vectors of extended goals, thus variables and their evaluated coefficients, are the systemic wellbeing functions as the objective criterion conveying the overarching functional ontology of unity of knowledge.

Some Economic Calculations in Tawhidi and Shari'ah Perspectives: Contrasting Examples

The evolutionary learning composition of Tawhidi wellbeing model with its functional ontology and associated circular causation relations in endogenous inter-variable complexity contrasted with the linearity of shari'ah perspectives results in different forms of definitions and calculations. The basis of the difference in meanings and calculations is the pervasive presence of organic relational unity of knowledge according to Tawhid as law; and separable way of shari'ah calculations in respect of its premise of methodological linear independence between variables (goals, as in maqasid al-shari'ah). The Tawhidi methodological worldview premises all calculations on the principle of pervasive complementarities, hence inter-variable and multidimensional participation between goals in their extended forms across knowledge, and knowledge-induced space and time dimensions. Shari'ah methodology legitimizes calculations on the basis of marginal substitution between variables or methodological independence between them. According to these two contending views of wellbeing (maslaha), the maqasid-methodology emulates the neoclassical economic worldview and social Darwinism of marginal substitution between variables, hence goals. These two worldviews are pervasive in the two opposite domains of reasoning, while globally interpreted the neoclassical and social Darwinist views are ingrained in competition and substitution and postulates of rationalism leading to economic rationality and socio-scientific dialectics of conflict, competition, and methodological individualism. Thus, cooperation as organic relational ethics of unity of being and becoming is denied (Minogue, 1963; Buchanan, 1999).

Two brief examples can bring out this contrasting distinctiveness in economic calculation: gross domestic product and inflation.

Gross Domestic Product Measured by Tawhidi Methodological Approach

Gross Domestic Product Measured by Tawhidi Methodological Approach in Terms of Micro-foundation

The Tawhidi law of unity of knowledge between the good things of life while avoiding the rejected things as codified by the commonality of choices between the Qur'an and maqasid al-shari'ah-at-Tawhid grounds all of economic reasoning on the micro-foundations and on the micro-foundation of macroeconomic theory. Thereby, the wellbeing function is always in respect of these two ground works of economic theory in Tawhidi perspectives, while invoking the primal ontological methodology of monotheistic worldview in unity of knowledge as the identification of Tawhid as law.

The life-fulfilment choices of goods and services $\{\mathbf{x}(\theta)\}$, with prices $\{\mathbf{p}(\theta)\}$ in the ethicizing market economy induced by unity of knowledge 'θ' between the good things of life, yields the life-fulfilment vector, $\mathbf{y}(\theta) = \{x_1(\theta), x_2(\theta), \ldots, x_n(\theta), p_1(\theta), p_2(\theta), \ldots, p_n(\theta)\}$. In the Tawhidi methodological context, expression (2.3) and thereby expression (2.4) are determined as wellbeing in terms of the $\mathbf{y}(\theta)$-vector. The inter-variable endogenous causal relations of unity of knowledge are derived as,

$$\begin{aligned} x_i(\theta) &= f_j\left(x_j(\theta), \mathbf{p}(\theta), \theta\right), \quad i = 1,2,\ldots,n; j = 1,2,\ldots,n; i \neq j. \\ p_i(\theta) &= f_j\left(p_j(\theta), \mathbf{x}(\theta), \theta\right), \end{aligned} \tag{2.5}$$

$\theta = F(\mathbf{y}(\theta))$ as the quantitative form of the wellbeing function (maslaha function) in endogenous variables of ethicized market transformation of life-fulfilment goods and services.

Imam Ghazali (Karim, n.d.) and Imam Shatibi's baskets of life-fulfilment goods and services are thus extended by endogenous force of consciousness 'θ' inducing a wide valuation of the maslaha functions across evolutionary learning systems. Rawls (1971) also wrote on his Difference Principle of A Theory of Justice in terms of his primary goods of social wellbeing. This was true of Schumacher. Tobin's idea of measure of economic welfare rejected calculation of bads from GDP, like environmental pollution, health hazards, and crime and punishment costs, and we could include cultural differences between the industrialized countries and the developing countries. The whole system of Human Development

Index has rejected the macroeconomic way of calculating GDP by the national income expenditure and income accounting to allow for social inclusion (HDI)

After the evaluation of the system of circular causation relations, the values of $\{p_i x_i\}$ are evaluated. Thereby, aggregation in the Tawhidi context is determined by, $(p.x)[\theta] = \prod_{i=1}^{n}(p_i x_i)[\theta]$ (Klein, 1946, "Remarks on the Theory of Aggregation", *Econometrica*, 14:4, pp. 303–312). This expression denotes the gross domestic product as the geometric aggregate of the values of life-fulfilment goods and services in terms of microeconomic foundation of market process induced by knowledge values 'θ'.

GDP Measured by Tawhidi Methodological Approach in Respect of Micro-foundation of Macroeconomics

If $\{x_i(\theta)\}$ denote variables like GDP ($Y(\theta)$), Consumption Expenditure ($C(\theta)$), Investment Expenditure ($I(\theta)$), Government Expenditure $G(\theta)$, and Net Trade ($NX(\theta)$), all of which are endogenous in nature, the principle of life-fulfilment choices and ethical market premise of calculation of GDP prevails. Following equations apply for inter-causality between the variables of the vector $\mathbf{Z}(\theta) = \{Y, C, I, G, NX\}[\theta]$. Note that G-variables, which are exogenous in the mainstream case of macroeconomics, become micro-foundational of macroeconomics in Tawhidi sense due to the participatory nature of organic interrelations between the variables. Government and institutions therefore play their participatory roles in decision-making and moral/ethical choices. The system of equations can be evaluated as in the case of expression (2.5). The value of GDP ($Y(\theta)$) can be calculated in terms of the other variables and vice versa. The micro-foundation of macroeconomic variables in $\mathbf{Z}(\theta)$ is implied by the existence of the wellbeing function in the $\mathbf{Z}(\theta)$-vector, subject to the inter-causal relational equations. This feature of the evaluative system implies the existence of pervasive endogeneity by complementarities (participation) between the variables.

Inflation Measured by Tawhidi Methodological Approach

From case (i) given above, the average price of a basket of life-fulfilment goods and services with evaluated outputs and prices is given by,

$$P(\theta) = \sqrt[1/n]{\prod_{i=1}^{n} p_i(\theta)}. \quad (2.6)$$

Thereby, the rate of change of the average price level is given by, $P(\theta)^\wedge = (1/n)\left[\sum_{i=1}^{n} p_i(\theta)^\wedge\right]$. Symbol '^' denotes rate of change in prices. This represents the meaning and measure of inflation.

The formal expression of all functional relations in the case of circular causation and the quantitative measurable expression of wellbeing function in the geometric product form of expression (2.6) readily yield the meaning of the evaluated coefficients. The values of the evaluated coefficients by estimation and simulation give the elasticity measures of the dependent endogenous variable in relation to the other system of endogenous variables in the circular causation variables. Such elasticity measures provide perspectives of decision-making in policy formulation, programme development, and academic visions of change and continuity.[8]

The other case of weighted arithmetic average of the prices cannot be used, because in that case the presence of 'θ' will be nullified due to the absence of relative prices (inter-price dependency) in the measurement of

[8] In a general case of formulation of product expressions such as $\log y(\theta) = \sum_{i-1}^{n} \alpha_i(\theta) \times \log x_i(\theta)$, with all variables and coefficients of n-number of variables induced by parameters of degrees of unity of knowledge signified by $\{\theta_i\}$, the elasticity coefficients are given by, $\partial \log y(\theta)/\partial \log x_i(\theta)$ = partial elasticity, $\epsilon_{y,xi} = \alpha_i(\theta)$. The meaning of the partial elasticity coefficient is 'what is the percentage sensitivity of 1% change in $x_i(\theta)$ on $y(\theta)$, across the whole system of circular causation relations between the variables of the θ-induced vector, $\{y, x_1, x_2, \ldots, x_n\theta\}[\theta]$'. The inter-variable sensitivity impacts of θ-parameter are derived from the above expression by writing as follows as induced by 'θ',

$$d\log y(\theta)/d\log\theta = \sum_{i=1}^{n}\left[\partial \log y(\theta)/\partial \log x_i(\theta)\right] \times \left[d\log x_i(\theta)/d\log\theta\right]; \text{i.e. } \epsilon_{y,\theta}$$

$$= \sum_{i=1}^{n}\left[\epsilon_{y,xi}\right] \times \left[\epsilon_{xi,\theta}\right] = \sum_{i=1}^{n}\left\{\alpha_{y,xi} \times \epsilon_{xi,\theta}\right\}[\theta]$$

as the expression of total elasticity of the dependent variable to the critical θ-induction in terms of the Tawhidi methodology of unity of knowledge. These kinds of elasticity expressions are not readily available with linear additive forms of circular causation equations. Consequently, the Tawhidi sensitivity impacts of inter-variable endogenous complementarities cannot be explained on methodological grounds. The resulting utilitarian forms, despite enforcing ethical parameters, cannot explain the continuously interactive, integrative, and evolutionary (IIE) complexity nature of complementary forms.

inflation rate. Because shari'ah could not direct human creativity towards such Tawhidi methodological thought, the latter days' proponents of 'shari'ah-compliance' rendered their subservience to mainstream economic and socio-scientific methods. The ideas surrounding inflation and price theory could not be discovered by shari'ah in respect of the inter-variable complementarities in the wellbeing function of life-fulfilment development choices. Consequently, θ does not induce the variables in the light of Tawhidi methodology of organic unity of knowledge. The average price level is not focused on the bundle of life-fulfilment goods and services in participative development regime. A very vast extent of differences result in respect of price theory of value and utilitarian theory of utility and welfare contrary to Tawhidi objective criterion of wellbeing (maslaha) and its various concomitant concepts. We examine these now.

Definitions and Concepts in Analytical Treatment by Tawhidi Methodological Worldview Contra Shari'ah

Usul (Fiqh)

In traditional Islamic literature, the maqasid al-shari'ah is traced to its origin in Islamic jurisprudence. This traditional reasoning to derive the tenets of shari'ah from Qur'an and sunnah makes the argument that the Qur'an is limited in the edicts of jurisprudence, there being less than only 620 verses referring to matters of law (Kamali, 2003, p. 25), and the rest to matters of belief and morality. Thus, between shari'ah and the Qur'an, despite the claim that the shari'ah arises from the Qur'an, there are common references to law, rules, and choices on recommended and avoidance on contrary beliefs and acts of life. Where such common references exist and they are in agreement with the sunnah, and are mutually discoursed in the Qur'an, the claim of shasri'ah to refer to the Qur'an and sunnah as the common foundations of the origin of Islamic law is justified. But there remains the great majority of more than nine-tenths of total verses (6200 ayaths) that are derived by human reasoning on revelation. This approach to understanding the Qur'an and sunnah and deriving the primal ontological foundations (methodology) in the Qur'an is subject to possible vagaries and influences of space, time, and variations of interpretations. The history of shari'ah is indeed ruptured to such external influences of changes in Islamic civilization, dynasties, and cultural and spatial factors (Hooker, 1984; Constitutional Rights Foundation, "The origins of Islamic Law", http://www.crf-usa.org/america-responds-to-terrorism/the-origins-of-islamic-law.html, visited Oct. 13, 2018).

Subsequent to this anomaly in determining the commonness and differentiation in methodological foundations of Islamic jurisprudence in the Qur'an, shari'ah could not establish the substantive meaning of the ontological, epistemological, and phenomenological foundation of the Qur'an as the completeness of Islamic law. Thereby, all collateral parts of shari'ah, namely ijtihad, qiyas, istihsan, juristic fatwa (see below), and maslaha are subject to methodological difficulties. These have continued to rest and contrast within their traditional domains. Yet such vagaries of the law are not in the nature of the over-encompassing holism of Tawhid as law according to Qur'an (48:23)[9]

Tawhid

In the perspective of Tawhid as the true, universal, and unique Qur'anic law of Islam, the discursive foundation of deriving exegesis, rules, and application from its source (primal ontology) of organic unity of knowledge is premised on the dynamics of shura combined with tasbih. In this way alone the Qur'anic phenomenology gets induced in the worldly order and assumes its generality and details of organic unity of being and becoming. A dynamic knowledge-induced systemic worldview of unity of knowledge in 'everything' emerges. The errors of differentiation in interpreting rules and erecting opinions (ra'y and fatawa) at the level of methodological individualism caused by independence of religious sects (madhabs) are reduced and changed into complementary and participative unity of interpretation based on organic unity of knowledge as the sure rule of Tawhid as law reflecting its invincible power of monotheistic unity in conscious and continuous sustainable systems of 'everything'.

Shari'ah

Shari'ah despite its claim of being law or codification (maqasid al-shari'ah) of the common space of Qur'anic law and itself has left out every functional representation of unity of knowledge that is found inherent in Tawhid concerning 'everything'. This great gap is evident in the

[9] Qur'an (48:23): "[This is] the established way of Allah which has occurred before. And never will you find in the way of Allah any change."

inadmissibility of shari'ah in all its forms and the fiqh that is ingrained in it, from its invoking the constellation of the cosmic whole and its distinctive methods premised on the world and the details of cosmic entirety. This missing gap exists and deepens in reasoning and revelation, according to shari'ah, in spite of the inseparable causality between the heavens and the earth, and with all that is embedded under the earth and above the earth. Such a precept of a priori reality of monotheism that remains endogenously inseparable from the a posteriori reality by and is explainable and extendible forms the methodology of systemic unity of knowledge. It is not to be found in shari'ah in any of its forms or reasoning concerning revelation and reasoning. Thus, Tawhid as law and methodology of monotheistic unity of knowledge is missed out by maqasid al-shari'ah and its various derived forms.

Usul Al-Fiqh

Usul al-Fiqh means the premise of jurisprudence. Yet, as we have noted above from historical evidences that, this juristic background despite its claimed reliance on the Qur'an and sunnah lacks serious drawback in methodological completeness on ontological grounds. Thereby as well, the epistemological and phenomenological groundwork of truly Tawhidi methodology of unity of knowledge in 'everything' suffers of the Qur'anic epistemic worldview. Consequently, both the academic and the applied functions of maqasid al-shari'ah are lacking in establishing the interdependent legal, moral, and belief aspects of all the Qur'anic verses taken together. That otherwise would convey the embedded monotheistic inter-causal meaning of inter-causal complementarity. The codification of the maslaha (wellbeing) and the mafasid (harm) choices remains incomplete and inadequately determined in the absence of qualifying such choices on the basis of Tawhidi methodological worldview of unity of knowledge and the implications on the unified world-system that so emerges.[10] Also refer to Qur'an (2:216) on the matter of exact codification by the Qur'an as opposed to this experience by human vacillation as by rationalist reasoning.[11] This is a mistake found in the shari'ah claim of equivalence

[10] Qur'an (3:60): "The truth is from your Lord, so do not be among the doubters."

[11] Qur'an (2:216): "Fighting has been enjoined upon you while it is hateful to you. But perhaps you hate a thing and it is good for you; and perhaps you love a thing and it is bad for you. And Allah Knows, while you know not."

(simultaneity) between revelation and reasoning. Rather the primal causation in this case is from revelation to knowledge of unity of being and becoming as revelation precedes creation as world-system.[12]

Tawhid as Law

The moment Allah decreed the creation with the command, "Be, and it is", Tawhid as law was bestowed to the world-system of 'everything'. This overarches the totality that forms the generalized system, and the infinitesimal details (specifics, particularities) as multiverses of the full universe. The hidden events that preceded along the divine order of revelation of the Qur'an comprises Tanzeel (divine descent) as revelation brought to bear on the heart and mind of Prophet Muhammad from the Preserved Tablet called Lauh Mahfuz and revealed in the Blessed Night of the month of Ramadhan. Secondly, the entire Qur'an was revealed by Allah's command through Angel Jibreel over a span of 23 years in accordance with various events across the 23 years history in the life of Prophet Muhammad and in references to history as it has developed and will evolve over time. The Qur'an centres its emphasis on its philosophy of history. When the Qur'an narrates events according to its central worldview of Tawhid as law, it is to support its philosophy of history across knowledge, space, and time dimensions. At the very moment that Allah commanded revelation to its state of Being and Becoming at the moment of 'Be, and it is',

Tawhid had transcended from its primal origin into its corporeal functioning governing 'everything'. Tawhid could thereby be unravelled by concrescence of moral and material law centred in the monotheistic principle of unity of knowledge and its codification by socio-scientific explanation in the order of 'everything'. On the other hand, Tawhid in its divine descent assumes the metaphysical meaning of attributes of Allah. Yet such attributes establish the primal ontology of Allah's oneness represented as functional law in the order of the nature and functioning of 'everything'. Thus, while the metaphysics of Tawhid remains beyond corporeality,[13] the corporeal functioning of Tawhid at the moment of Being establishes Tawhid as the law. Its most distinctive worldview is premised on monotheistic unity of knowledge and its representation in respect of the recon-

[12] Qur'an (2:117): "Originator of the heavens and the earth. When He decrees a matter, He only says to it, 'Be, and it is.'"

[13] Qur'an (22:74): "No just estimate have they made of Allah: for Allah is He Who is strong and able to Carry out His Will. …"

struction of the world-system of 'everything' in terms of the cardinal ontology of unity of knowledge.

Shari'ah as Rule Setting

From the above-mentioned explanations on the nature of usul al-fiqh in its claim to cover what its scholars assert to be the shortage of juristic verses in the Qur'an, it is implied that the codification of rules done by shari'ah in all its forms is a mix of common verses and human objectified rules. The danger in regards to determining the Tawhidi groundwork of Qur'anic law now lies not in its learned discursive approach, as by the mujtahids (ulalamr). Rather, the serious concern with shari'ah codification remains with the absence of a well-defined methodology as the primal ontology that unravels the foundational precept of monotheistic unity of knowledge and the knowledge-induced world-system (a'lameen). The substantive exegesis of the Qur'anic verses is thus missed out. That is, to originate knowledge and reasoning with the ontological foundation of Tawhid as law, and then if needed to invoke the discourse of the learned community.[14] Yet, all of these sequential processes must be embodied in the essence, formalism, and application of monotheistic oneness as explained by Qur'anic organic unity of knowledge of inter-causal organism (Qur'an, 36:36). Such an approach adduces the singular nature of Tawhid as law as it becomes externalized in the being and becoming of the generality and particulars of all world-systems. The divine law of monotheistic oneness is thus actualized in the corporeal world-system as its law. Codification emanates from the well-definition and objectivity based on this most pronounced and singular precept of Tawhidi unity of knowledge. Shari'ah in all its forms has failed to understand this ontological origin as the methodology of law, while it takes account of human interpretations in the absence of this methodology as secondary mode of codification of derived rules. The resulting partitioning between substantive and procedural law on matters of codification is explained by Ghazali (trans. Karim, n.d.). See also Hart (1994).

[14] Qur'an (4:59): "O ye who believe! Obey Allah, and obey the Messenger, and those charged with authority among you. If ye differ in anything among yourselves, refer it to Allah and His Messenger, if ye do believe in Allah and the Last Day: That is best, and most suitable for final determination."

Reasoning (Ontology and Episteme)

According to Tawhid as law, which externalizes the divine monotheistic essence into the corporeal world-system of 'everything' ever after the Qur'anic event of "Be, and it is", revelation came first and thereby arose the corporeality of reason. Thus, opposed to the shari'ah understanding, there remains the continuous flaw that reason and revelation are coterminous consequences. This mistake is so deeply ingrained in all aspects of shari'ah that it has left scars of Qur'anic judgement on worldly implications arising from the divinely moral-material functioning of Tawhid as law.

Examples of this mistake can be noted as follows: The complete equivalence between Tawhid as law and maqasid al-shari'ah is unworkable as long as the unity of organic interrelations between all parts of the Qur'an is displaced as the primal methodological episteme of monotheistic oneness externalized in the order of meta-reality. Hence, from the very inception on methodological grounds there cannot be equivalence between Tawhid as law and maqasid al-shari'ah as codification. Such codification, instead of being able for determination on the primacy of Tawhid, rests on the incomplete domain of its five elements. These are, namely, protection of religion, protection of reason, protection of life, protection of progeny, and protection of property and property rights. Even in this rudimentary state of incompletely realizing the wide expanse of moral reality, the elements remain as pillars in their methodological independence, thus barring the organismic relations of unity of circular causations. The result then is a misunderstanding of the meaning of morality and ethics according to the Tawhidi law of organic unity of inter-variable relations. The resulting absence of systemic worldview in shari'ah has forever remained stillborn contrary to the universal system-oriented worldview of unity of knowledge of the Qur'an.[15]

As another example, the problem of what comes first—chicken or the egg—in respect of revelation and reason, is exhibited in the mundane example of the functioning of money in Islam. In Tawhidi methodological perspective, the identity problem between the quantity of money and prices in the equation of exchange, $Mv = p.y$ taken up with the Money-Commodity-Money (MCM)-causal equations[16] raises a problem of

[15] Qur'an (13:1–4).

[16] Quantity of money identity in the MCM-causal relations: $M1 \Rightarrow C1 \Rightarrow M2 \Rightarrow C2 \Rightarrow M3 \Rightarrow$ continuity. Shari'ah literature has failed to explain how money converts to more or less money ($M \Rightarrow M$) through its commodification ($\Rightarrow C$) except to render this monetary process

indeterminism: M as quantity of money, v as velocity of money circulation, p as prices of goods and services over money circulates (transactions), and y as real output of goods and services.

The argumentative flaw in the expression revelation-reason complementarity is further established by the Qur'an (76:1–3).[17] This verse establishes the primordial status of monotheistic knowledge in Tawhid prior to the creation of reason as artefact in human knowledge. In accordance with the primordial nature of divine knowledge premised in Tawhid followed thereafter by reason, there are two aspects of the shari'ah failure based on the concept of revelation-reason complementarity. Firstly, if reason is conjoint with the a posterior world of sensate forms, then reason is totally human generated. The monotheistic origin of Tawhidi unity of knowledge in the Qur'an is fully abandoned. Secondly, if revelation is assumed by a priori realm of divinely commanded moral precepts without a functional link with the a posteriori realm of the world-system, then the primordial functional nature of Tawhid as law as at the moment of 'Be, and it is', is negated. The over-encompassing monotheistic law of creation has no relevance. Then the Qur'anic worldview of the world-system (a'lameen) along with its signs of Allah (ayath Allah) explaining the monotheistic law in the totality of creation (sunnat Allah) are all contradicted. Thirdly, the belief on complementarity of revelation and reason, which must occur in the a priori realm of Tawhid as law contradicts the primacy of monotheistic

as being illegal as interest-bearing function, wherein fiat money becomes bank savings as commodification (C). In Tawhidi relational worldview of unity of inter-causal elements, knowledge plays the singularly most primal role. Thereby, the above MCM expression of monetary and commodity inter-causal relationship causes financial instruments to be the medium of transacting a quantity of money into money. Now all the variables in the vector $\mathbf{z}(\theta) = (M, C, M, F, p, y)[\theta]$ become endogenously interrelational by the medium and functioning of knowledge in respect of inter-variable unity of organic being and becoming. Now the price of transaction in Qur'anic sense is riba-free; rather only distinct between maslaha and mafasid, according to how M is mobilized in the ethicized market of exchange in the good things of life, $\{y(\theta)\}$. Thus, in every part of the MCM model the conversion of money into money is well-defined by the market operation on its determining of price and output. This is the Tawhidi approach because of the fundamental functioning of $\{\theta\}$ on all the variables. It is not comprehended by shari'ah due to its absence of the organic relational methodology.

[17] Qur'an (76:1–3): "Has there [not] come upon man a period of time when he was not a thing [even] mentioned?[1] Indeed, We created man from a sperm-drop mixture that We may try him; and We made him hearing and seeing [2]. Indeed, We guided him to the way, be he grateful or be he ungrateful.[3]."

knowledge and its embedding in the created world-system by subtle mapping caused by sunnah.[18]

In terms of each of the above points of fallacy of arguments in shari'ah to invoke it as the origin of Islamic law or codification of law, the usul al-shari'ah imitates Kantian mechanics of reasoning by knowledge. This approach to the theory of knowledge is the intrinsic nature of the human perception in rationalism. The Kantian epistemic similarity of reasoning in shari'ah is a deeply methodological issue that is of a different nature than the utterance of divine invoking by shari'ah. Kant as well desired for an embedded relationship between his conception of moral imperative in a priori reasoning and a posteriori reasoning in material world-system. But this desire belied him on methodological grounds of the absence of the subtle mapping to carry the moral imperative of a priori to the material order of a posteriori except by the medium of human reasoning and intervention alone. In this regard, Kant wrote (Kant, 1949, p. 25; reprinted 1949, p. 25): "In what follows, therefore, we shall understand by *a priori* knowledge, no knowledge independent of this or that experience, but knowledge absolutely independent of all experience. Opposed to it is empirical knowledge, which is knowledge possible only *a posteriori*, that is, through experience."

The idea of rationalism was thus born. It denied divine origin to reasoning. Thereby the distinctive parallel realities of reasoning emerged between rationalism and Tawhid as the ontological origin of unity of knowledge and the opposite rationalist pluralism of heteronomous reasoning (Bhaskar, 1994). In regards to Tawhid as the foundation of unity of knowledge, the Qur'an (51:11) declares regarding such contest between Tawhid as law and rationalism as vagaries of human perception: "Those who (flounder) heedless in a flood of confusion."

Reasoning according to Tawhid as law of monotheistic unity of knowledge while encompassing its origination in the primordial attribute of the ultimate truth of the oneness of Allah comprises continuous process of derivation and evolution of unity of knowledge and its unification of the world-system. Reason is thereby ordained by the primordial ontology of Tawhid in its totality and which is carried forward by the sunnah on to the formalism of concrescence of ideas pertaining to the generality and

[18] Qur'an (6:101): "To Him is due the primal origin of the heavens and the earth: How can He have a son when He hath no consort? He created all things, and He hath full knowledge of all things."

particulars of the world-system. Out of formalism governed by the embedding of evolutionary knowledge concerning unity of being and becoming there arises the unified design of the world-system. The emergent design is thereafter constructed in material form of its enduring wellbeing brought from the ontological origin, formalism, and into the moral and material order of the world-system. This is the goal of wellbeing called maslaha as the reasoned worldview of unity of knowledge and of the knowledge-induced unified and evolutionary world-system in the midst of learning. The indefinite continuity of evolutionary knowledge of Tawhidi consilience in the moral and material embedded reality bears the truth of reason born of the primal ontology of Tawhid and its permanence.

Contrary to such a design of organic unity of the evolutionary learning world-system, there is the opposite of differentiation, methodological independence, and methodological individualism. We refer to this opposite worldview as 'de-knowledge'. It bears the mark of rationalism and thereby of the differentiated nature of reason and reasoning arising from the 'de-knowledge' premise of rationalism. While shari'ah by itself does not premise its intent on 'de-knowledge' as the nature of rationalism, yet the critical absence of a distinctive methodology of monotheistic unity of knowledge, around which all inquiries relating to moral and socio-scientific objectivity and applications can be based, causes delusion regarding the primacy of Tawhid as law. The maqasid al-shari'ah understood from Ghazali's viewpoint rested its origin on Tawhid and Tawakkil (trust in Allah), and redemption of revelation as primal over and above rationalism. Yet the methodological depth of the all-comprehensive law of 'everything' between the heavens and the earth as the precept of Qur'anic law endowed in Tawhid is never to be found. Thereby, although Ghazali referred to causality as a process of evolving knowledge, it could not emerge as a most subtle dynamics of the participatory organic universe of inter-causal relations of unity of knowledge. The subsequent shari'ah scholars could not realize this distinctive overarching nature of Tawhid as law, and shari'ah within it. Only lately Auda (2008) attempted to assign the system characteristic to the study of maqasid al-shari'ah, yet failing to include a dynamic system approach within the completeness of Tawhid as law. In the end, the study of Tawhid as law of the heavens and earth remained absent among all generations of Muslims following Ghazali (Buchman, 1998). Thus, avoiding the contradistinction of Tawhid as the ultimate law, the Qur'anic verse (Qur'an 45:18) declares: "Then We put you, [O Muhammad], on an ordained way concerning the matter [of religion]; so follow it and do not follow the inclinations of those who do not know." It is self-pronounced by any who would otherwise associate the meaning of the 'ordained way' with reason and reasoning comprising a posteriori or the dysfunctional a priori nature of shari'ah in its notion of simultaneity between revelation and reason. The Prophet Muhammad was a mortal on whom Allah bestowed the last and complete revelation; as the Qur'an (42:52–53) declares: "And thus We have revealed to you an inspiration of Our command. You did not know what is the Book or [what is] faith, but We have made it a light by which We guide whom We will of Our servants. And indeed, [O Muhammad], you guide to a straight path. The path of Allah, to whom belongs whatever is in the heavens and whatever is on the earth. Unquestionably, to Allah do [all] matters evolve." In spite of all the anomalies in the understanding and exegeses of these Qur'anic verses, there are indeed common grounds between the Qur'an and shari'ah. These common grounds remain sustained between Tawhid as law and shari'ah as codification of Tawhidi law. But beyond these commonalities, shari'ah and its fiqh remain as human determined out of complicated derivations of rules that can be questioned.

By being unable to comprehend the vista of supercardinal law and its derived meaning by evolutionary learning centred on the Tawhidi-type methodology of unity of knowledge, shari'ah as bounded text of codification of rules has failed to substantively give meaning to the organic unity of knowledge regarding the heavens and the earth and all in between. In practical stance the derived principle of pervasive complementarities between the good things of life has failed in shari'ah and its applications. This problem emanating from the axiom of scarcity in economic rationality has caused a blind immersion of Islamic unitary worldview into mainstream economics and social sciences.

The primal ontology of Tawhid has remained as belief but dysfunctional as dynamics of creation in 'everything' in shari'ah. This scholarly implication is similar to the Kantian characterization of heteronomous differentiation between a priori and a posteriori. Contrarily, reason as material artefact of creative essence is born by the primordial command of a priori as Tawhid in its total meaning, acting on creation in a posteriori. The continuous interconnectivity between a priori and a posteriori, thereby between revelation and reason, is explained by the sustainability of the evolutionary learning process, [Qur'an = {Revelation → Sunnah → Knowledge} ↔ (Reason ↔ world - system) → continuity to hereafter].

Polity-Market Interaction

The absence of the methodological dynamics of participatory unity between transdisciplinary diversity of things in usul al-shari'ah constructed on the basis of usul al-fiqh implies how presently marginal substitution born out of scarcity of resources in the axiom of economic rationality has deepened in the so-called 'Islamic economics, finance, and social inquiry'. Consequently, several major examples can be seen to exist as aberration of the Tawhidi law of unity of knowledge. For instance, presently, 'Islamic economics and finance', by being absorbed in mainstream economic doctrines, has not been able to frame Islamic epistemic thought on the basis of dynamic preferences caused by evolutionary learning of unity of knowledge. Consequently, on the problem of distribution of resources between economic efficiency (growth) and social justice (fairness), the great trade-off between these is absorbed in Islamic economic, finance, and social thought by the mistaken idea of priority while rejecting pervasive complementarity between these two good things of life, namely economic growth and efficiency and simultaneously social justice. The issue of trade-off as marginal rate of substitution in rational economic choice theory needs attention. Likewise, the idea of priority is much echoed but least understood in 'Islamic economics' except as a journalistic and not a technical terminology (Al-Qardhawi, 1990).[19]

[19] Al-Qardhawi, Y. (1990). "Priorities of the Muslim Movement in the coming phase", lecture in Masaryk University, Doha, Qatar. https://is.muni.cz/el/1421/podzim2013/RLB379/um/45655867/45656045/Yusof_Al-Qardhawi_-_s.72_-_83.pdf On the technical meaning of priority that does not violate the meaning of pervasive complementarities between the good things of life in an evolutionary learning phase of Tawhidi law applied to 'everything': Consider two phases of resource allocation that yield the following quantities of goods and services: $-(x_1, x_2)[\theta]_1 = (5, 10)[\theta]_1$; $(x_1, x_2)[\theta]_2 = (7, 15)[\theta]_2$. $dx_1/dx_2 = 1/5 > 0$, implying simultaneous increase, hence complementarity between $x_1(\theta)$ and $x_2(\theta)$ between phase 1 and phase 2. This complementarity is shown on shifting resource allocations caused by 'θ'-effect as in the case of continuous technological change. The rate of change in $x_1(\theta)$, $dx_i(\theta)/x_1(\theta) = 2/5$ is lower than the rate of change in $x_2(\theta)$, $dx_2(\theta)/x_2(\theta) = 5/10$. Thus, there is priority of $x_2(\theta)$ over $x_1(\theta)$, but not substitution between these allocations along evolutionary learning resource allocation trajectories induced by $\{\theta\}$-values. Thus, pervasive complementarities of economic expansion between the two goods prevail by continuity of 'θ'-values. Multiple goods can be similarly treated. Contrary to this phenomenon, differentiating groupings of goods by way of methodological individualism between such groupings permanently grounds rational choice theory and the meaning of scarcity of resources in marginal substitution concept of scarcity, competition, and methodological individualism (Quinton, 1989; Yolles, 1998).

The commodity specification of complementarity contrary to the economic rationality meaning of marginal rate of substitution leads to subtle difference and consequences in respect of polity-market preference behaviour. This topic of dynamic preferences plays a substantive role in the institutional decision-making. We explain this now to bring one more significant difference in Tawhidi way of deriving polity-market organic unity of relations that is not found in the mainstream formation of 'Islamic economics, finance, and social theory'.

As a result of the utilitarian derivation of shari'ah rules and decision-making based on such derived rules (ahkam), there are imminent meanings of shura as discursive consultation; ijtihad as authentic meanings derived by authoritative findings by consensus on matters arising from the Qur'an and sunnah; qiyas as debated approach to deriving rules by usul al-fiqh via the exercise of analogical deduction. This raged as cantankerous matter that laid bare a great expanse of disagreement between the logicians and rationalists, as of the Mutazilites and Asharites on the one hand against the theological madhabs, especially of Hambal, Hanafi, Shafei (Hallaq, Wael B. (2005, *The Origins and Evolution of Islamic Law*, p. 125). There is also the topic of ijma as consensus in consultation (shura) and juristic preference called istihsan and in qiyas. To the extent that istihsan is a social preference formation to attain degrees of wellbeing by choices of possibilities, it is an important approach towards attaining such maslaha and avoiding mafasid, and continuing on the experience of relating all the emanating practices and wellbeing consequences to Tawhidi law of unity of knowledge. While this worldview is critical for Tawhid as law of 'everything', it is not noticed to function within shari'ah. A methodological reason for such dysfunctioning in shari'ah of the interactive, interactive, and evolutionary learning processes in continuum over meta-science in knowledge, space, and time dimensions is the absence of compound, nonlinear, and complex aggregation of relations. Methodological permanence of unity of knowledge takes its expressions within this kind of dynamic preferences. In the case of shari'ah, the implication of its utilitarian formalism results in linear utilitarian aggregation (Harsanyi, 1955).

Quinton, A. (1989). *Utilitarian Ethics.* La Salle, IL: Open Court.

Yolles, M. (1998). "A cybernetic exploration of methodological complementarism", *Kybernetes: International Journal of Systems and Cybernetics*, 27:5, pp. 527–542.

A formal expression of dynamic preferences in polity-market interaction contrary to the utilitarian approach that abounds in shari'ah without having a creative outcome is explained in the following way:

Let $\wp_i(\theta)$ denote preferences for ith decision-making unit; i = 1 (market), 2 (institution), 3 (morality). The nature and degree of organic intercausal relationship between the three subsystems is parameterized by 'θ'-values denoting the average complementing knowledge value between the stated three subsystems. Decision-making and organic complementarities in a shared inter-systemic condition involve interaction as discourse leading to integration as consensus, which then is followed by evolutionary learning to further continuously enhance knowledge towards betterment of the discursive issue for gaining on wellbeing (maslaha) and reducing harm as mafasid ('de-knowledge). These elements of a continuous evolutionary learning process towards decision-making using knowledge-induced preferences are embedded in the following expression: $[\cup^{\text{interaction}}\cap^{\text{integration}}\{\wp i\}][\theta] = \wp(\theta)$ as the unified complementary preference function of collective decision-making.

This simulation effect of θ-embedding of discursively determined preference function for shared and collective decision-making is imminent in the Tawhidi methodological worldview of its consultative decision-making involving institutional and system-oriented organic unity of knowledge. Such a dynamic discursive engagement is referred to in the Qur'an (42:38) as shura.[20] The subsystems that engage in agential and institutional interactions, integration, evolutionary learning (IIE), and their continuity of processes are referred to in the Qur'an as acts of worship in exchange of involving the monotheistic law and the generality and details of the issues under study (tasbih).[21] Thus, inculcating consciousness in decision-making is a part of the unity between tasbih and shura concerning 'everything'. We refer the inner essence of conscious decision-making in 'everything' encompassing the heavens and the earth as shura-tasbih.

[20] Qur'an (42:38): "And those who have responded to their lord and established prayer and whose affair is [determined by] consultation among themselves, and from what We have provided them, they spend."

[21] Qur'an (22:18): "Do you not see that to Allah prostrates whoever is in the heavens and whoever is on the earth and the sun, the moon, the stars, the mountains, the trees, the moving creatures and many of the people? But upon many the punishment has been justified. And he whom Allah humiliates—for him there is no bestower of honor. Indeed, Allah does what He wills."

The complexity and non-linearity born of shura-tasbih cannot be derived from the linear pattern of utilitarianism. In the linear case, the social preference caused by the three subsystems mentioned above causes, $\wp = \sum_{i=1}^{3} \wp_i$. In this case, the methodological independence between the three subsystemic preferences causes cessation of the IIE-learning process in continuum. Hence, 'θ' is either invariant or is nullified. Or of it changes, it would be in divergent dialectical phases. Each of these processes shows the continuity of competition by way of methodological independence and methodological individualism between the generality and details of the various subsystems.

Shari'ah failed to ascribe decision-making to inter-causal unity of knowledge in complex and non-linear processes by its gurus. The critical example of such a ready acceptance of degenerate processes, models, and their inferences based on linear spaces without the IIE-learning processes is found in the use of econometric models blindly to linear cases of inter-variable relations. They are devoid of the Tawhidi law. Consequently, as well, the maslaha model has remained linear in form, and thus devoid of the knowledge of organic circular causation relations. The so-called 'Islamic economics, finance, and social' inquiries have therefore failed to render their original resilience with its Tawhidi ontological episteme leading to its derived phenomenology in continuum. A robust critical heterodox theory of transdiciplinary intellection has not emerged under shari'ah to enable Muslim mind to critical thinking over a long time in the history of intellectualism. This is equally true of the failure of shari'ah in its linear non-creative thinking to provide a robust functional theory of shura-tasbih as evolutionary learning process and all refined thought that this thinking otherwise would render to the world of learning.

Tawhid Contra Shari'ah in Respect to Essentials of Qur'anic Law

Terms and concepts like 'usul', 'revelation and reason', 'qiyas', 'shura', 'istihsan', and 'maslaha' (therefore mafasid) acquire distinctive concepts and meanings in the Tawhidi worldview contrary to those of shari'ah. The principal grounds on which this contrariness occurs is the reference to all terminologies and precepts in abstracto-empirical conceptions in the Qur'anic worldview to be based on the monotheistic organic unity of knowledge that Tawhid most distinctively emulates. This is the central coordinate around which all functioning concerning the Qur'anic bestowing in Tawhid as law revolves. The reflection of inter-causal relations of unity of knowledge between the blessed and recommended possibilities prescribed by the Qur'an and their codification as good choices while avoiding the forbidden ones is the surest explication of the principles of pervasive complementarities between life-fulfilment acquisitions. In shari'ah, the codification of the blessed and recommended things of life (halal at-tayabah) is not based on the precept of unity of knowledge. Rather, the mistaken concept of priority in choices of the good things of life prevails. This is a mark of scarcity of resources that arises from the occidental thought regarding socio-scientific rationality. This furthermore arises from the mistaken conceptions of complementarity between revelation and reason. In turn these precepts are included in rationalism. The next missing attribute of shari'ah is the absence in it of any precise methodological premise that derives and imparts the Tawhidi foundation of the Qur'an and sunnah on to the study of world-system. Instead, the meaning of usul while belonging to the Qur'an and sunnah are deeply linked to sectarian (madhab) origin of variant Muslim schools of thought and practices. This has been the way with the historical development of shari'ah. The singular exception in this regard was with the Tawhidi basis of developing maqasid al-shari'ah in the conceptual ways according to Ghazali, yet not in a methodological way emulating the Tawhidi methodological worldview. The ontology as usul of shari'ah thus remains an Islamicized form of rationalism.

The above two points of the primal ontology of Tawhid remaining the keynote of the Tawhidi methodological worldview, which shari'ah has failed to invoke in its usul al-fiqh, the Tawhidi worldview of the critical terminologies rests on the singular axiom of an evolutionary learning postulate with interaction and integration to define the underlying principle of monotheistic unity of knowledge. Thereby, terminologies of usul, istihsan, qiyas, complementary between revelation and reason, shura, maslaha, and the like as in shari'ah are all subjected to the unique premise of organic participation between ideas, intra-systems and inter-systems in evolutionary epistemology, formalism in the form of dynamic relations, and phenomenological application of Tawhidi methodology in 'everything' in knowledge, space, and time continuity. All reason and intellection emanate from Allah and return back to Allah in completeness. This means the absolute completeness of Tawhid as law in the universe of absolute reality (Qur'an).[22]

[22] Qur'an (2:156): "Surely we are Allah's and to Him we shall surely return."

While the goal of maslaha is derived from Tawhidi methodology of organic unity of knowledge working out in its completeness of formal usul, balance, and conscious bestowing to the world-system in its entirety, so also the Qur'an declares the nature of falsehood and forbidden, mafasid. Yet the methodology and method of Tawhid as the ultimate law prevails uniquely on both maslaha and mafasid in terms of the opposite meanings of truth and falsehood, respectively. The uniqueness of Tawhidi methodology can be proved in the following way: The primal ontology of Tawhid as the embodiment of the Qur'an is identically established in the Hereafter. Thereby, the universe of 'everything' is self-contained in itself, through the evolutionary and cumulative learning process of the world-system in accepting truth and rejecting falsehood. Thereby, the entire universe is distinctly disjoint between truth and falsehood. Any undecidable between truth and falsehood is due to incompleteness of knowledge and is finally determined as truth or falsehood with the advance of knowledge of unity, $\{\theta\}$, as truth. 'De-knowledge' as the variant mathematical complementation in opposite to truth characterizes differentiation, competition, and their coterminous choices. If all of knowledge of truth, including in it the truth manifestation of falsehood completes 'everything', say Ω, thereby, the disjoint union of the multiverse of truth, say $T(\theta)$, and of falsehood, say $F(\theta)$, completes the fullness of the universe of 'everything'. That is, $T(\theta) \cup F\{\theta\} = \Omega$, with $T(\theta) \cap F(\theta) = \phi$. Falsehood thus means the opposite meaning of truth in Ω. In conclusion, any indeterminacy and misnomer of terminologies differently from its Tawhidi roots is in contradiction to the Qur'an.

Conclusion: Tawhid Contra Shari'ah in Their Referring to the Qur'an and Sunnah

Such misunderstanding either results from incompleteness of knowledge, as is the case of a gap of learning in shari'ah to source its origin in the full Tawhidi embodiment of Qur'an. The other way is its containment in the rationalist ego of man. This defies any Qur'anic link of Tawhid as law. This is the case of rationalism and is not truth. Ibn al-Arabi wrote in regards to falsehood stemming from rationalism (Chittick, 1989): "Useless knowledge is that which is disconnected from its source and origin, i.e. from the divine reality. Any knowledge outside of Tawhid leads away from Allah, not toward Him. But knowledge within the context of Tawhid allows its possessor to grasp the interconnectedness of all things through a vast web whose centre is the divine. All existent things come from Allah and go back to Him." The freeing of shari'ah from its rationalist Islamicized nature is urgently needed in order to equivalence maqasid al-shari'ah fully with Tawhid as law of unity of knowledge and the imminent methodology. When this happens, the true law in Islam of 'everything' will be Tawhid as law and none else. All other realms of knowledge and learning will remain comparative and contrasting in nature for critical examination. Ghazali noted about this blessed experience of knowledge, that which is singularly derived from Tawhid as the primal ontology of monotheism and the generality and details of the world-system: He wrote (trans. Karim, n.d., p. 245): "When knowledge was puffed up in his heart, his oil was enkindled. Then light upon light came to him. Then knowledge said to him: Value this moment greatly. Open your eyes, so that you may find the path. When he opened his eyes, he found the pen of God as described. It is not made of reed, it has got no head. It is incessantly writing in the mind or soul of men. He said being surprised at it: What a good thing is knowledge."

A critically important part of the methodological worldview of Tawhid as law and of shari'ah is the analysis of the topic of completeness of the law as the universal and unique one extending on matters of life beyond life. The world-system in its extended meaning comprehends such a kind of holism without finitude, yet not infinitude; for total reality from the Beginning of Creation to the End as continuity in Hereafter must have the same primal ontological origin followed by evolutionary epistemology, and continuing in sustainability by the phenomenology of ontic phenomenology based on the ontological and epistemic foundations. These all together emanate from the unique source of the Qur'an and sunnah, the latter as the explained mapping of the Qur'an on to the total phenomenology of the world-system.[23] The Qur'anic meaning of phenomenology is

[23] The meaning of the term 'phenomenology' throughout this work stands for the study of uman consciousness arising from certain primal ontological and then carried by epistemic use into the ontic nature of functional (engineering) ontology of being as corporeal objects. The entire analytical details of such objects are thus encapsulated in the being and nature of 'everything' under study. A most critical property of the systems of objects by their generality and details both intra-systems and inter-systems is conveyed and the idea of ontologically and epistemologically derived phenomenology as idea by inter-causal relations. The relations so derived and studied can be of the IIE-learning processes as in and by Tawhid as law; or it can be of the differentiating dialectical type but explained uniquely by Tawhid as law. The analytical validity of the moral substance of being and through its becoming as analytical derivation

thereby different from the occidental one as raised by occidental philosophers Heidegger, Husserl, and others contrary to Imam Ghazali's Qur'anic explanation of mind and the universe (Badawi, 2002)

Yet there is a similarity between the two methodological ideas on phenomenology. Stanford Encyclopedia (op cit) writes: "Thus, phenomenology leads from conscious experience into conditions that help to give experience its intentionality. Traditional phenomenology has focused on subjective, practical, and social conditions of experience. Recent philosophy of mind, however, has focused especially on the neural substrate of experience, on how conscious experience and mental representation or intentionality are grounded in brain activity."

The above explanation of the field of consciousness pervades and arises from the Tawhidi law in relation to the inherent unity of knowledge between permissible things as codified while avoiding the forbidden ones; and deriving the unique and universal methodology that abides in respect of this primal ontology of 'everything'. It conveys a methodological worldview that overarches from the Beginning of Creation to the Total Reality of Creation in the End in Hereafter. Nothing is remised in Tawhid as divine reality and its corporeal meaning as law. This Qur'anic meaning of the Beginning and End needs subtle explanation in respect of the finite multiverses and the largest world-system extending to the Hereafter.

The briefest expression of such a knowledge-induced universal inter-causal and inter-temporal relationship is explained by the expression, Tawhid $\rightarrow$ knowledge - induced world - system $\rightarrow$ Tawhid in both the finite sequences of the learning subsystems and in the largest scale of the Hereafter. Thereby, in the largest scale the symbiotic relationship is expressed by, Tawhid $\leftrightarrow$ Tawhid $\Rightarrow$ Beginning $\equiv$ End and the End $\equiv$ Beginning identically. Fullness of Tawhidi unity of knowledge is completed in Tawhid. Inter-causality ends at the two ends of completed total reality. The sensate world-system yields the final and new system called the Hereafter in Heaven and Hell through the Great event of Doomsday (Kiyamah).[24]

is thus unbarred on the conscious distinction between Truth (unity of being and becoming) and Falsehood (differentiation of being and becoming of objects and cognition). The complete methodological worldview is thus established by the IIE-learning processes unravelling over Primal Ontology, Epistemology, Functional Ontology of Logical Formalism, Phenomenology, and Continuity in Knowledge, Space, and Time Dimensions.

[24] Qur'an (24:37): "Their (only) fear is for the Day when hearts and eyes will be transformed (in a world wholly new)."

Tawhid in its meaning of total monotheistic reality and as the corporeal law explaining 'everything' at the moment of the divine command, "Be and it is", bestows distinctive meaning to the world-system in its completeness from the Beginning and End. Tawhid is thereby the true and complete law. Its uniqueness and universality in this sense, as summarized above, cannot therefore be supplanted by any other law, not even shari'ah in any of its forms. At best the clarion call 'shari'ah' in any methodological attempt is an utterance rather than substantive meaning and essence of Qur'anic law. The conclusion of this chapter is that there remains a greatly impending gap to reunite maqasid al-shari'ah and all its dispensations with Tawhid as law, and thereby to emulate the methodological worldview of unity of knowledge in the transdisciplinary fold. When this worldview is actualized, even though in the specialized though scholarly venue, then maqasid al-shari'ah will attain its true Qur'anic status as maqasid al-shari'ah al-Tawhid.

References

Al-Qardhawi, Y. (1990). Priorities of the Muslim Movement in the Coming Phase. Lecture in Masaryk University, Doha, Qatar. Retrieved from https://is.muni.cz/el/1421/podzim2013/RLB379/um/45655867/45656045/Yusof_Al-Qardhawi_-_s.72_-_83.pdf

Arbouna, M. B. (2007). The Combination of Contracts in Shariah: A Possible Mechanism for Product Development in Islamic Banking and Finance. *Thunderbird International Business Review, 49*(3), 341–369.

Auda, J. (2008). *Maqasid as Shari'ah as Philosophy of Islamic Law* (p. 45). Herndon, VA: The International Institute of Islamic Thought.

Badawi, M. (2002). Phenomenology. In *Stanford Encyclopedia of Philosophy*. Stanford University.

Bhaskar, R. (1994). *Critical Realism*. London, UK: Verso.

Buchanan, J. M. (1999). The Domain of Constitutional Economics. In *The Collected Works of James M. Buchanan* (Vol. 1). Indianapolis, IN: Liberty Fund.

Buchman, D. (1998). *The Niche of Lights* (p. xiii). Provo, UT: Brigham Young University Press.

Chittick, W. C. (1989). *Sufi Path of Knowledge*. Albany, NY: State University of New York Press.

Ghazali, I. in Karim (op cit, n.d.).

Hallaq, W. (2005). *The Origins and Evolution of Islamic Law*. Cambridge, UK: Cambridge University Press.

Harsanyi, J. C. (1955). Cardinal Welfare, Individualistic Ethics, and Interpersonal Comparisons of Utility. *Journal of Political Economy, 63*, 309–321.
Hart, H. L. A. (1994). *The Concept of Law* (p. 92). Oxford, UK: Clarendon Press.
Haykel, B. (2014). The Political Failure of Islamic Law. *Occasional Papers*. Yale Law School Legal Scholarship Repository.
Hooker, M. B. (1984). *Islamic Law in South-East Asia*. Oxford, UK: Oxford University Press.
Kamali, M. H. (2003). *Principles of Islamic Jurisprudence*. Cambridge, UK: Islamic Text Society.
Kant, I. (1949). Critique of Pure Reason. In C. J. Friedrich (Ed.), *The Philosophy of Kant*. New York, NY: Modern Library.
Karim, F. (n.d.). *Imam Ghazzali's IhyaUlum-Id-Din*. Lahore, Pakistan: Shah Muhammad Ashraf.
Klein, L. (1946). Remarks on the Theory of Aggregation. *Econometrica, 14*(4), 303–312.
Lawson, T. (2009). *Ontology and Economics: Tony Lawson and His Critics*. London, UK: Routledge.
Mahdi, M. (1964). *Ibn Khaldun's Philosophy of History*. Chicago, IL: The University of Chicago Press.
Minogue, K. (1963). The Moral Character of Liberalism. In *The Liberal Mind* (pp. 166–174). Indianapolis, IN: Liberty Fund.
Quinton, A. (1989). *Utilitarian Ethics*. La Salle, IL: Open Court.
Rawls, J. (1971). *A Theory of Justice*. Cambridge, MA: Harvard University Press.
Rida, R. (1983). Introduction. In *Al-Mughni* (Vol. 1). Beirut, Lebanon: Dar al-Kitab al-Arabi.
Wallerstein, I. (1998). Spacetime as the Basis of Knowledge. In O. F. Borda (Ed.), *People's Participation, Challenges Ahead* (pp. 43–62). New York: The Apex Press.

CHAPTER 3

Casting Maqasid Al-Shari'ah in Tawhidi Methodological Worldview

Tawhid, Maqasid al-Shari'ah in Relation to the Precept of Ummah

Using the premise of the maqasid al-shari'ah, Islamic scholars have worked in the concept of the ummah, the world-nation of Islam. Yet the idea never arose from the Tawhidi epistemological foundation. It was thus not possible for the Islamic learned scholars to understand the global epistemological context of what the Qur'an meant by the meaning of the ummat al-wasatah—the balanced nation. The Qur'an carries on to declare that this overarching intellection of the epistemological meaning of the balanced world-nation of Islam as the momentous event.[1] In this chapter, we interpret the meaning of the balanced nation and the momentous event of the ummah as the overarching experience of soul, mind, and matter. This specification too is derived from the Qur'an.[2] The meaning of the ummah thus remained for long that of the nation states blended together by the common religion of Islam, yet not understanding the Tawhidi unifying meaning of the ummah as the experience in soul, mind, and matter universe of the Qur'anic all-fathoming world-system, the a'lameen.

Consequently, the understanding of what came to be known in recent times as Islamic economics and finance and Islamic science, remained bereft of the Tawhidi epistemological moorings (Guessoum, 2011; Nasr,

[1] Qur'an (Chapter Bakarah on ummah).
[2] Qur'an, Fussilat, Signs of Allah.

M. A. Choudhury, *Tawhid and Shari'ah*,
https://doi.org/10.1007/978-3-030-49087-4_3

1978). Islamic scholars are led into a quagmire of confusion in finding nothing new and contributing nothing new in the otherwise ought-to-be rich field that could be truly Islamic economics and finance. The intellectual predicament arose from the Islamic scholars to copy mainstream economic reasoning and models that rest upon rationalistic means with a palliative of the Islamic morality and ethics. But along the same lines the great economists have written on such issues (Keynes, 1963; Coase on Smith, 1994; Walras, 1898). Yet they were not Islamic economists. The fundamental roots of reasoning in Islamic economics, which cannot be any other than the monotheistic semblance of organic unification in a systemic sense of organic interrelations between the variables and entities of maqasid al-shari'ah, was never invoked. Consequently, Islamic economic and finance scholars forgot the epistemological beginning of a scientific reasoning. This kind of scientific orientation was what Albert Einstein wrote to his friend Neils Bohr (1985).

Likewise, the understanding of science as process premised on Tawhidi unity of knowledge and unity of the world-system never appeared to the Islamic scientists (Iqbal, 1934). Islamic scientists were thereby unable to understand the nature of the evolutionary learning universe in the midst of Tawhidi unity of knowledge as otherwise Wallerstein (1998) and Heisenberg and Anshen (1958) ventured out to study consciousness in the space–time structure of micro sub-universes. Likewise, by missing out the overarching presence of Tawhid as law in the scheme and order of the multiverse problems in the dimensions of knowledge, space, and time could not emerge out of the limited domain of space–time structure without the embedded nature of morals as knowledge. In the end, we have inherited the maqasid al-shari'ah and its extensions not as a general-system study of the interrelated multiverse in the midst of the Tawhidi understanding of the unified world-system resonating the corporeal meaning of unity of being. Maqasid al-shari'ah thus remained a partial study of the universe, limited to worldly affairs in the light of a set of interpretations called fiqh. Fiqh instead of continuous revisiting the Qur'an and the sunnah to regenerate knowledge parted away from the development of understanding of the Tawhidi law and the world-system. This was never the intent of the majestic maqasid al-shari'ah or of the medium of interpretation, fiqh, if the Qur'an and sunnah remained continuously as the starting point regarding Tawhid and the knowledge-centred world-system.

When further modern issues arose, as the problem of sustainability and ecology, the shari'ah scholars failed to understand the nature of such problems as synergistic evolutionary learning in the midst of an ever-widening balance of complementary relations between the maqasid al-shari'ah variables and entities in the light of Tawhid as law. Sustainability degenerated into a sustainable development issue. The result was a narrow understanding of sustainability as sustainable development as depreciable capital.

In the post-modern era, the Islamic scholars find their understanding of the shari'ah to be limited to answering certain intellectual issues as diverse specifics. It is for these human discursive failures that Weeramantry (2001) and Hassan point out the overloaded nature of the shari'ah as the study of legal contracts (uqud). The general-system study of Tawhid in the midst of a wider field of holistic valuation has been missed out. The cause and effect of this remiss was the abandonment by the Islamic scholars of the Tawhidi methodological worldview that prevails by unity of knowledge to explain the unifying universe in its generality and specifics across the dimensions of knowledge, space, and time.

In the post-modern era, it is pressing to address the problems faced by humanity and the global ecological, social, economic, and financial order by the recasting of the maqasid al-shari'ah along the framework of the Tawhidi methodological worldview. This is to carry the Tawhidi primal ontology, epistemology, and phenomenology to the realm of substantive methodology and its applications.

Objective

The objective of this chapter is to address the general-system perspective of Tawhid as law in terms of its objective and purpose and in the light of the central epistemological core of monotheistic oneness of knowledge and unity of the induced world-system taken up in its generality and specifics. By the wellbeing objective function and its formalism within the generalized system approach to the study of the Tawhid contra shari'ah, we invoke the following question raised for example by Henderson (1999, p. 56). She puts the contrasting picture between wellbeing and capitalist globalization succinctly:

> The good news is that this (capitalist globalization) is forcing us to "go inside ourselves" and ask some pretty basic questions: What do I *want* to pay attention to? Who am I and what do I want written on my tombstone? Such basic defensive reactions will define the growing sectors of our Attention

> Economies and their inexorable shift from material goods (measured by traditional GNP/GDP per capita), to services and more intangible factors in living standards, measured by the new Quality of Life scorecards.

Our objective in this chapter is to answer the above-mentioned type of inherent question regarding the future of the human race in relation to the cosmic entirety. That is because we will argue via formalism that sustainability of our common future lies on evolutionary learning in the light of unity of knowledge and unity between entities of *all* world-systems. In this book, we will formalize the model arising from Tawhidi methodological worldview in respect of the monotheistic episteme of unity of knowledge and its induction of the organic and relationally unified world-system. We will then study specific example emanating from the generalized treatment of selected issues and problems of economics, finance, science, and society by invoking the Tawhidi methodological worldview contra their study by applying shari'ah.

The thesis we want to propound is that, with the new wave of thinking that post-modernity has brought with it, there is a parallel and powerful rise of a unitary and symbiotic way of thinking and organizing life that emanates solely and wholly from the Islamic worldview of Tawhid as law.

A Brief Review of the Literature

Mankind today is facing global entropy. The principal cause of this is the construction of a fallen social edifice of thought, behavior, and institutions that distract individuals and collectivities away from the moral and ethical foundations of a good and shared social order.

On this matter, Korten (1996, p. 11) writes:

> The continued quest for economic growth as the organizing principle of public policy is accelerating the breakdown of the ecosystem's regenerative capacities and the social fabric that sustains human community; at the same time, it is intensifying the competition for resources between rich and poor—a competition that the poor invariably lose.

Heilbroner presents his dark views on the human prospect (1991, p. 11):

> There is a question in the air, more sensed than seen, like the invisible approach of a distant storm, a question that I would hesitate to ask aloud did

> I not believe it existed unvoiced in the minds of many: "Is there hope for man?"

On the specter of the economic future of mankind, Keynes wrote (1963, p. 319):

> I may do well to remind you, in conclusion, that the fiercest contests and the most deeply felt divisions of opinion are likely to be waged in the coming years not around technical questions, where the arguments on either side are mainly economic, but around those which, for want of better words, may be called psychological or, perhaps, moral.

On the grand design of unity of understanding reality between self and the cosmological order, there is the beautiful quote from Kant (transl. Friedrich, 1949):

> Two things fill the mind with ever new and increasing awe and admiration the more frequently and continuously reflection is occupied with them; the starred heaven above me and the moral law within me. I ought not to seek either outside my field of vision, as though they were either shrouded in obscurity or were visionary. I see them confronting me and link them immediately with the consciousness of my existence.

On the critique of the shari'ah being narrowed down to the study of legal contracts in the hands of latter Muslim communities, Weeramantry (2001, pp. 118–20) has commented. "Much of the tenets of the Islamic Law became an imposition on the individual by the state for adherence to, rather than a change in the will, conduct and sustained attitudes of Muslims as self-governing behavior on ethical preferences." We argue in this book that such a compulsive attitude of the state was due to the failure of the traditionalists (fuqaha) in understanding the endogenous learning dynamics of the ethicality concept in a general-system study of Tawhid as law concerning the world-system.

The failure on this front of developing the true Islamic worldview as a general-system study of Tawhidi unity of knowledge and its induction of the generality and details of the world-system was noted even by Ibn Khaldun. On the matter of shari'ah, Ibn Khaldun considered this law to be an ideal, yet not a law practiced during his time (Mahdi, 1964).

Hassan (2002) points out a clear absence of a general-system approach to understanding the synergistic relationship between the different holistic parts of the entire world-system. The traditionalists developed the shari'ah along narrow lines of legal contracts (uqud) in different and segmented areas. Thereby, a generalized universal theory of multiverse relations could not be brought into shari'ah, which otherwise is the keynote of Tawhid.

A similar ambivalence of maqasid al-shari'ah is found in Doi's work, which has not even mentioned the great Tawhidi purpose and thus has not explicitly referred to the Tawhidi foundation of shari'ah. The continued confinement with the traditionalists of shari'ah to the limited domain of muamalat (worldly affairs) and personal affairs has severed the great intellection of the maqasid al-shari'ah from the domain of the divine law of monotheistic oneness concerning the world-system of all diversity (sunnat Allah). Doi (op cit, p. 11) writes in this regard: the basic principles of shari'ah are pointed out to be (1) interest of society; (2) emphasis on relieving hardship over promoting benefit; and (3) preference for priority as a means of exigency of needs. All these points totally dissociate the maqasid al-shari'ah from the broad goals as pointed out by the great Islamic jurist, Imam Malik. This important principle of extending the shari'ah is pointed out by Doi (op cit, p. 8). That is, wellbeing for the public purpose ought to extend the bounds of the shari'ah to include such matters that do not belong to the traditional rules of the shari'ah confined to muamalat.

Hossain (2003) provides a tacit definition of the field of shari'ah, which defines the conflicting usage of this field, its meaning, and application. This explanation shows the narrowed-down scope of the shari'ah to a theocratic understanding of society and worldly affairs, rather than integrating the comprehensive study of science and the social order in details that comprehend the overarching field of intellection according to the Qur'anic law of monotheism, Tawhid.

Hossain writes (p. 13):

> The *shari'ah* provisions were compiled by the *ulama* (Islamic scholars) during the Middle Ages using the discipline of *fiqh* (Islamic jurisprudence), *hadith* (Prophet Muhammad's sayings), *ijma* (consultation and consensus) and *qiyas* (analogy).

Such limitation of the domain of overarching intellection of Tawhid has been the actions of the traditionalists for a long time now. The situation fuelled the incessant conflict between the traditionalists and the modernists in understanding the realm of the res extensa and res cogitans as reflecting the holism of the divine law premised solely on Tawhid as law.

In reference to the grand design of the unitary multiverse and the Tawhidi intellection within this domain, Gulen (2006, pp. 148–49) writes:

> We use 'the horizon of hope' to mean traveling beyond the visible dimension of existence, and considering existence as an interrelated whole in the absence of which things and events cannot be perceived as they really are. Nor can its essence and relation with the Creator as well as the relation between them and humanity be grasped. Scientific disciplines that conduct their own discourse largely in isolation from one another and the prevailing materialistic nature of science that has compartmentalized existence and life cannot discover the reality of things, existence, or life.

In search of the general-system construction of the universe as ontological reality in the light of its objective and purpose, we now enter a functional ontology of logical formalism derived from the Tawhidi epistemological roots and implicated in the construction of relationally unified world-system. The elements of this new regime of moral transformation form momentous normative and positive questions. They embrace momentous changes in the ideals, praxis, and enactment of the vision of a globally acceptable ethical and moral worldview that must supersede all issues of life that global exchange and partnership embrace for common wellbeing.

In the post-modern understanding of the greater symbiotic world-system following the advance of challenging issues, it is well-known that legal theory has extended into medical questions of stem cell, abortion and pro-life, and euthanasia. Legal theory has overarched into questions including evolutionary theory in education curriculum and environmental ethics. The day is not too far when mathematical logic could be brought to bear on legal arguments. On such issues of epistemic analysis of language and sentences, see Shatibi's (Masud, 1994) ideas and in recent time Godel's (1965). Presently DNA is being allowed in legal evidences. Such legal developments overarch into new and challenging domains of socio-scientific questions that are distinctly transdisciplinary in nature.

In such areas, like any other legal theory, shari'ah should have much in transcending to explain the socio-scientific reality. Such possibilities to offer answers to the great purpose and objective of existence can be explained by linking the origin of maqasid al-shari'ah in the Qur'an, the sunnah, and the authoritative rules derived by epistemic investigation, to Tawhid as the primal ontology, the essential law. Thereby, the great transdisciplinary horizon of Tawhid in its symbiotic nature of the paired multiverses would be required to lay down the true extended foundations of Tawhidi reality. This kind of an approach leads into the understanding of the generalized system context of including in it the open domain of the divine law, Sunnat Allah. Without such a true and potential extension, the shari'ah remains confined in the light of the Tawhidi methodological worldview.

Formalism of the Generalized Knowledge-Induced World-System in the Light of Tawhid as Law

Meaning of Tawhid as Monotheistic Functional Ontology

The meaning of Tawhid in the Qur'an is more than simply the primordial belief on the Oneness of God. Rather the all-comprehending meaning of Tawhid embraces the primordial belief in divine Oneness and the manifestation of this belief in the experiences of unity of the world-system as the domain of the Signs of God (Ayath Allah) in the order of 'everything'. Such a gapless overarching universal world-system by continuity of relationship across the res extensa and res cogitans of continua is referred to in the Qur'an as 'a'lameen'. The Qur'an further exhorts this conscious reflection of oneness in the heart, belief, and the order of 'everything' in the verse: "Al-Qur'an as-Zikr."

Thus, the maqasid al-shari'ah as the derivative of such Tawhidi reflective intellection across all things in its generalized system meaning is translated in the criterion of wellbeing as concept and measurement of balance reflecting organic oneness by unifying relationships. The primordial ontology is inseparably unified with the functional ontology of formalism, observation, and explanation. The deductive is united with the inductive reasoning. Traces of Kantian-type heteronomy (Choudhury, 2010) separating a priori reality from a posteriori reality is annulled and united together as continuity by circular causation to reason and understand reality in and across continua.

The Tawhidi epistemology is the learning process forming a methodological order of reasoning, thought, formal construction, and action (application), followed by inferences over continuity. This whole system of recursive learning is continuously regenerated in the knowledge, space, and time dimensions. Such a domain of 'everything' is premised on unity of knowledge and its unifying deductive-inductive circularity of reasoning as the very meaning of continuity res extensa and res cogitans across continuums (*Qur'an* 2: 115; Bakar, 1991).

To formalize, we define the Tawhidi episteme as the supercardinal topology, (Ω, S). Here, Ω denotes the Qur'anic super-space or supermanifold of knowledge (Dewitt, 1992). This yields only relations and responses but no corporeal limitation. The Qur'an declares in this regard, "Be and it is". S denotes Sunnah (guidance of Prophet Muhammad) as topological mapping of the Qur'an to living experience. Worldly knowledge including exegesis of Qur'anic verses, $\{\theta\}$, flows from (Ω, S). This is denoted by $\{\theta\} \in (\Omega, S)$. $\mathbf{x}(\theta)$ denotes the knowledge-induced vector, matrices, tensor, all pertaining to the problems under investigation. The vector $\{\theta, \mathbf{x}(\theta)\}$ spans the knowledge-induced world-system arising from the discursive experience of unity of knowledge of Tawhid as law.

Furthermore, $\{\theta, \mathbf{x}(\theta), t\}$ denotes the multidimensional triplet spanning the knowledge $\{\theta\}$, knowledge-induced space $\{\mathbf{x}(\theta)\}$, and time $\{t(\theta)\}$ dimensions. That is, the Tawhidi methodological worldview spans the following extensive and rich systems of interrelationships:

Tawhidi rich and complex world-system

Tawhid$\rightarrow_{(s)unnah}$ World-System $\rightarrow_{(s)unnah}$*Tawhid*: The large-scale closure
(Ω,S) → {0,x(0),t} → (Ω,S) through the passage of finitely learning world-System

↓

Islamic Embedded world-system as complex relational socio-scientific sub-system
It is characterized by an example

↓ (3.1)

Example

$E\{\cup_i \cap_j (R, \Upsilon, p; \text{Pref})[\theta]^s_{ij}$; R: resource—financial, natural, labor; Υ: output; p: price level;

i = interaction = 1, 2, …, n_1; j = integration = 1, 2, …, n_2; s = systems = 1, 2, …, n_3. This symbolism is an extension of Debreu (1959).

Furthermore, $dE/d\theta > \theta$, denoting evolutionary epistemology; $dE/d\theta' > \theta$, evolutionary 'de-knowledge' (rationalism).

Thus, the process of knowledge derivation from the Tawhidi epistemic core is represented by the inter-causal, inter-variables, intra-systemic, and inter-systemic Interaction (I), Integration (I), and Evolutionary (E), assuming the **IIE**-learning processes. IIE as process methodology forms the permanent property of the Tawhidi epistemological construction of pairing by evolutionary learning in unity of knowledge in 'everything' (Barrow, 1991). The Qur'an explains all this in its meaning of reorigination (khalq in-jadid) through the evolutionary learning processes of the world-system to its final end in the Great Event of the Hereafter (Akhira).

No quick and short, complete description can be given of such a comprehensive Tawhidi epistemological and methodological worldview. The imminent project is a vastly detailed methodological investigation. It carries a thorough analytical, applied, inferential, and policy and institutional implications (Choudhury, 2011).

A Generalized System Formalism of Unity of Divine Knowledge a la Tawhid as Law

The generalized system model of the Tawhidi unity of knowledge is now formalized to yield the foundation of the true Islamic law.

The epistemology of Tawhid is the principal law of unity of knowledge and its induction on the construction of the unified world-system. The Tawhidi epistemological premise comprises the totality of the Qur'an. We denote this by the symbol **Ω**.

The Qur'anic knowledge is carried by the medium of guidance of the Prophet Muhammad (Sunnah). We denote this transmitter mapping by S. S is the primal ontological mapping of the Qur'an into living experience.

Since the entirety of the Qur'anic knowledge is only a 'bit' in progressive sequences that can be derived from the divine law by *S*, symbolically, we denote this 'bit' of knowledge formation in the real world-system by the discursive composition of the holistic knowledge of **Ω** by *S* (thus (Ω, S)) to form the first category of knowledge by principally visiting (Ω, S). This primal discursive knowledge is denoted by the set $\{\theta*\}$.

The whole causation process, $[(\Omega, S) \rightarrow \{\theta*\}]$ comprises the generality of the primal ontology of Tawhid as law. From this origin, the 'bits' of knowledge flows yield the worldly discursive and constructive knowledge pertaining to details and specifics of issues and problems to be investigated. The result spans all world-systems in an evolutionary way across human potential and nexus of world-systems. $\{\theta\} \in [(\Omega, S) \rightarrow \{\theta *\}]$ is thereby the result of human comprehension of the world-systems by the mapping of a 'bit' of Ω onto the world-system by the medium of S. This phenomenon is established by the Qur'an, as Allah has bestowed only a 'bit' of His supercardinal knowledge on humans and creation, so as to learn therefrom of Tawhid.

The formation of $\{\theta\}$ is an extensively discursive process that incessantly evolves into higher levels of certainty and comprehension as the discursive process continues along the realm of investigation by means of the divine law. Such a discursive process as the enabling institution of application of Tawhid in the world-system is realized by the Qur'anic consultative institution called the shura in conjunction with the intricacies of the worshipping world-system (tasbih).

Shura is combined with authoritative derivation of rules by the learned ones in the Qur'an and the Sunnah, called ijtihad. The purpose and objective of the Shura combined with ijtihad is to search and discover by progressive learning the Tawhidi dynamics of the world-system. This form of worship of divine oneness in the form of explanatory socio-scientific relations obeying the organic pairing of unity is called tasbih. Thus, the knowledge derivation and development processes from the Tawhidi epistemic premise to its realization in the construction of the world-system in generality and details can be referred to as tasbih-shura evolutionary learning process It signifies the ontological, epistemological, and discoursed understanding of the worshipping world-system in terms of Tawhidi unity of knowledge.

We denote the 'bit' of the spontaneously unravelled knowledge flow $(\{\theta *\}, \{\theta\})$ on the ontological entities by the set $\{\mathbf{x}(\{\theta\})\}$, meaning that the emanation of the world-system arises spontaneously from the domain of the episteme of unity of knowledge through the impact of knowledge flows.

A constructed world-system at the end of the evolutionary learning experience in Tawhidi unity of knowledge is denoted by the spanning of the unified pairs over $(\{\theta\}, \{\mathbf{x}(\theta)\}, \{t(\theta)\})$. The organic pairing marks the *functional* ontology of the constructed world-system on the basis of the Tawhidi unity of knowledge. The formal process of deriving, studying, and evolving out of the learning multiverses in Tawhidi unity of knowledge comprise the logical formalism arising from the functional ontology of being and becoming of the Tawhidi knowledge-induced world-system.

The formalism following from the *functional* (different from the *primal*) ontology of the constructed world-system on the basis of unity of divine knowledge is evaluated by 'estimation' and 'simulation' in accordance with the guidance of the Tawhidi law. While the 'estimation' phase conveys the positive 'state of the investigated world 'as is', 'simulation' normatively constructs the 'investigated world as it ought to be'.

The formalism derived by functional ontology transforms, carried out by testing the degree of human achievement, response, and further actionable decisions sustains the evolutionary learning processes in Tawhidi unity of knowledge in continuity and continua of multiverse systems. We thereby move from the origin of the Tawhidi primal ontology comprising (Ω, S) to the derived formalism of $(\{\theta\}, \{\mathbf{x}(\theta)\}, \{t(\theta)\})$. $\{\mathbf{x}(.)\}$ is multidimensional along with $\{\theta\}$, as the evolutionary knowledge $\{\theta\}$ spans across diversely complementing nexus of world-systems. In our earlier terminology, such spanning world-systems comprise the grand and consciously centred ecological order. Thereafter, the world-systems learn and sustain themselves from the constructed ontology to higher evolutionary realms of the Tawhidi unitary worldview. That is, $(\{\theta\}, \{\mathbf{x}(\theta)\}, \{t(\theta)\})$ evolves into higher knowledge flows, and thereby into higher levels of the constructed world-systems.

The whole formalism from primal ontology via epistemology and functional ontology emanating from the Tawhidi epistemic origin is completed in the evaluation of the wellbeing function, subject to the circular causation relations between the selected variables and entities.

Such a formal explanation is explained below. The formal expression of evaluation of the wellbeing function, subject to circular causation relations, is written down as follows:

$(\Omega;S) \rightarrow \{\theta^*\} \rightarrow \{\theta\} \rightarrow (\{\theta\},\mathbf{x}(\{\theta\}))$ $\rightarrow$ evolutionary $\{\theta\} \rightarrow$ evolutionary $(\{\theta\},\mathbf{x}(\{\theta\})) \rightarrow$

$\downarrow$ Evaluation $\uparrow$
(comprising 'estimation'
Followed by 'simulation')
$W(\theta) = W(\mathbf{x}(\theta))$
Subject to circular causation of
Relations between variables as
Empirical sign of unity of organic
Relations in the light of the
*Tawhidi*methodological worldview
Of unity of knowledge:
$\mathbf{x}_i(\theta) = f_i(\mathbf{x}_j(\theta))$, $i \neq j = 1,2,\ldots,n$
and the expression of the measured
form of wellbeing in knowledge,
$\theta = F(\mathbf{x}(\theta))$, $\theta \in [(\Omega;S) \rightarrow \{\theta^*\}]$, the
domain of Tawhid as law. (3.2)

In expression (3.1), the portion,

$$\left[\{\theta\} \rightarrow \left(\{\theta\},\mathbf{x}\left(\{\theta\}\right)\right)\right] \subset \{(\Omega;S) \rightarrow \{\theta *\}\} \rightarrow \text{Evaluation}] \qquad (3.3)$$

marks a completed process of ontological construction of the world-system as derived from the epistemology of (Ω; S).

The portion,

$$\left[\text{evolutionary}\{\theta\} \rightarrow \text{evolutionary}\left(\{\theta\},\mathbf{x}\left(\{q\}\right)\right) \rightarrow \text{Evaluation}\right] \qquad (3.4)$$

marks an evolutionary phase of post-evaluated constructed ontology through the epistemic reference to Tawhidi unity of knowledge.

Two central issues need to be pointed out in expression (3.2). The medium of knowledge formation at any point along the sequences of evolutionary processes concerning the knowledge-induced world-systems is always of a discursive nature. Only the fundamental epistemology remains outside discourse because of its axiomatic nature in the constructed functional ontology. A discourse along a process is described by interaction leading to consensus (integration) followed by the beginning of a post-evaluated co-evolutionary process involving the tuple $(\{\theta\},\{\mathbf{x}(\theta)\},\{t(\theta)\})$ in the wellbeing objective criterion.

The important property of the Tawhidi String Relation (TSR) of expression (3.2) is circular causation that permanently defines the knowledge-induced process methodology. Processes are derived in a recursive manner under the force of the shura-tasbih—*ST*-experience. Sustainability of the processes in expression (3.1) is determined by the *ST*-experience, and thus by the evaluation of the wellbeing function, subject to circular causation between entities, variables and sub-relations of nexus in the unifying processes.

The third important point to note in the methodology of expressions (3.1) and (3.2) is its nature of openness and closure. The circular causation and continuous learning in the Tawhidi evolutionary learning experience being pervasive over unifying world-systems, these processes remain open. They are only temporarily bounded by evolutionary equilibriums that determine attained epochs of consensus (limiting values) in $(\{\theta\},\{\mathbf{x}(\theta)\},\{t(\theta)\})$ - values (Grandmont, 1989). But in the very large scale Tawhidi universe the completion of knowledge of *all* nexus of world-systems is towards Ω once again. Thus, while the component processes, and hence world-systems, remain open-ended in the Tawhidi universe, the very large-scale Tawhidi universe is closed though not bounded, for there is no corporeal meaning of the supercardinal nature of Ω. Only relational epistemology is meaningful.

The above explanation determines the interrelationship between the closed and open Tawhidi universes through co-evolutionary processes in the world-system as follows:

$$(\Omega;S) \rightarrow \text{Evolutionary}\ `\mathit{Alameen} \rightarrow \text{Evolutionary}\ `\mathit{Alameen}\ldots. \rightarrow \text{Hereafter} = \Omega \qquad (3.5)$$

Complex nexus interrelations in the $(\{\theta\},\{\mathbf{x}(\theta)\},\{t(\theta)\})$ - values and their recursive post-evaluation of simulated objective criterion occur as conceptually and empirically determined functional ontologies within this IIE-process methodology. The sustainability of the Tawhidi evolutionary learning experience is thereby pervasive in the unifying learning experience along the diverse trajectories of evolutionary equilibriums of such unifying world-systems. See elsewhere (Choudhury, 2000) for a mathematical and empirical formulation of the ensuing issues of expressions (3.2) and (3.5).

Conclusion: Reinterpreting Maqasid Al-Shari'ah in Light of Tawhidi Methodological Worldview

We begin this concluding section with a deduction from the socio-scientific conception of the generalized system model of Tawhid and the world-system. This is an original intellection that we have discussed above. It goes against the narrowed-down scope of the shari'ah to worldly (muamalat) matters only. The divide between these two viewpoints has a gamut of socio-scientific consequences. Besides the ones pointed out above between the traditionalists and the modernists on the endemic issues of non-universality and generalized system, it also opens up the rich nature of the Tawhidi domain.

This further vista of intellectual extension is over a vaster domain of diverse objectives, though the prevailing belief on merely five such objectives have been assigned to maqasid al-shari'ah. For instance, consider the following diversities within the prevailing ones):

Upholding the Islamic Religion

This category upholds the understanding of the Tawhidi epistemology in its holism of primal and functional ontology of conceptual and conscious real-world formalism.

Realizing and Upholding Justice

Justice is defined as balance of equitable relations between unitary forms of interacting variables and entities; and extending the precept of social justice from the human world to the domains of interrelations between the human world and the rest of the world-system (Bayrakli, 1992).

Protecting Human Reason (aql)

This category comprises differentiation between the epistemic meanings of reason, rational, and rationalism in Tawhidi socio-scientific thinking; static and intertemporal reasoning on issues of decision-making and predictability concerning resource allocation; sustainability and evolutionary learning on the basis of Tawhid; aql formed by the phenomenology of Tawhidi unity of knowledge as episteme.

Upholding the Values of Family

Family is treated as social miniscule (Choudhury, 1996). Family is treated as the seat of human resource development; and family is studied in terms of intertemporal social security.

Protecting Property Rights

Property rights is treated in terms of intergenerational shared property rights. Thereby, valuation of property rights follows the right to acquire and own; giving allowance for contractual variations in the cases of variations by intertemporal variations of exigencies. Property rights and social rights establish God-centred organizations and properties.

From our diversification of the traditional five-point maqasid al-shari'ah, it can be seen that the resulting variation focuses on the integrated linkages exemplifying Qur'anic pairing as the sign of Tawhidi unity of knowledge between the good things of life in the generalized system representation across knowledge, space, and time dimensions. We will now bring out this epistemological and functional ontological capability in respect of specific issues of economics, science, and society in the light of the extended version of maqasid al-shari'ah first guided by the primal ontology of Tawhid as law.

References

Bakar, O. (1991). *Tawhid and Science*. Penang: Secretariat for Islamic Philosophy and Science. The Science University of Malaysia & Nurin Enterprise.

Barrow, J. D. (1991). *Theories of Everything, the Quest for Ultimate Explanation*. Oxford: Oxford University Press.

Bayrakli, B. (1992). The Concept of Justice (*Adl*) in the Philosophy of Al-Farabi. *Hamdard Islamicus, V*, 3.

Bohr, N. (1985). Discussions with Einstein on Epistemological Issues. In H. Folse (Ed.), *The Philosophy of Niels Bohr: The Framework of Complementarity*. Amsterdam: North Holland Physics Publishing.

Choudhury, M. A. (1996). A Theory of Social Systems: Family and Ecology as Examples. *Kybernetes: International Journal of Systems and Cybernetics, 25*(5), 21–38.

Choudhury, M. A. (2000). The Qur'anic Model of Knowledge: The Interactive, Integrative and Evolutionary Process. In *The Islamic Worldview, Socio-Scientific Perspectives* (pp. 41–73). London: Kegan Paul International.

Choudhury, M. A. (2010). Unity of Knowledge Versus Kant's Heteronomy with a Reference to the Problem of Money, Finance and Real Economy Relations in a New Global Financial Architecture. *International Journal of Social Economics, 37*(10), 764–778.

Choudhury, M. A. (2011). Dynamics of the Shari'ah and the Islamic World-System. *King Abdulaziz University Journals: Islamic Economics, 23*, 1.

Coase, R. H. (1994). Adam Smith's View of Man. In *Essays on Economics and Economists*. Chicago, IL: The University of Chicago Press.

Debreu, G. (1959). *Theory of Value, an Axiomatic Analysis of Economic Equilibrium*. New York, NY: John Wiley.

Dewitt, B. (1992). *Supermanifolds*. Cambridge: Cambridge University Press.

Godel, K. (1965). On Formally Undecidable Propositions of Principia Mathematica and Related Systems. In M. Davies (Ed.), *The Undecidable*. New York: Raven Books.

Grandmont, G.-M. (1989). Temporary Equilibrium. In J. Eatwell, M. Milgate, & P. Newman (Eds.), *New Palgrave: General Equilibrium* (pp. 297–304). New York, NY: W.W. Norton.

Guessoum, N. (2011). *Islam's Quantum Question*. London: I.B. Tauris.

Gulen, M. F. (2006). *Toward Global Civilization of Love & Tolerance*. Somerset, NJ: The Light, Inc.
Hassan, H. (2002). Contracts in Islamic Law: The Principles of Commutative Justice and Liberality. *Journal of Islamic Studies, 13*(3), 257–297.
Heilbroner, R. L. (1991). *An Inquiry Into the Human Prospect*. New York, NY: W.W. Norton.
Heisenberg, W., & Anshen, R. N. (Eds.). (1958). *Physics and Philosophy*. New York, NY: Harper &Brothers Publishers.
Henderson, H. (1999). *Beyond Globalization, Shaping a Sustainable Global Economy*. New York, NY: Kumarian Press.
Hossain, M. Z. (2003). *Global Islamic Politics* (p. 13). New York, NY: Longman.
Iqbal, M. (1934). *The Reconstruction of Religious Thought in Islam*. Oxford: Oxford University Press.
Kant, I., & Friedrich, C. J. (Eds.). (1949). Critique of Pure Reason. In *The Philosophy of Kant*. New York, NY: Modern Library.
Keynes, J. M. (1963). The End of Laissez Faire. In *Essays in Persuasion*. New York, NY: W.W. Norton.
Korten, D. C. (1996). *When Corporations Rule the World*. London: Earthscan.
Mahdi, M. (1964). *Ibn Khaldun's Philosophy of History*. Chicago, IL: The University of Chicago Press.
Masud, M. K. (1994). In Islamic Research Institute, International Islamic University (Ed.), *Shatibi's Theory of Meaning*. Islamabad.
Nasr, S. H. (1978). *An Introduction to Islamic Cosmological Doctrines*. Boulder, CO: Shambhala.
Wallerstein, I. (1998). Spacetime as the Basis of Knowledge. In O. F. Borda (Ed.), *People's Participation, Challenges Ahead* (pp. 43–62). New York: The APEX Press.
Walras, L. (1898). *Etudes d'Economie Sociale*. Lausanne: F. Rouge.
Weeramantry, C. G. (2001). *Islamic Jurisprudence, an International Perspective*. Kuala Lumpur: The Other Press.

CHAPTER 4

Tawhid Contra Shari'ah Treatment of Qur'anic Worldview

Qur'anic Terms Unclear in Terms of Shari'ah

The definitions of moral values, despite being the high watermark of the Qur'an and the sunnah, remain imprecise. An example of such imprecise understanding of moral values is of justice, fairness, equity, and wellbeing. Authors like Kamali (2005) wrote on this issue from the juristic and shari'ah points of view in a way as if the moral values as given by the Qur'an are proscribed. Consequently, the absence of precise definitions associated with specific moral values makes it fuzzy to establish individual, institutional, and social character identification and policies involving such specific values. The result then is institutional cost and benefit derailment on the legal and regulatory implications of social policy and the understanding of precise shari'ah contexts. An example on the other hand, regarding the specific definition of justice in western philosophical context, is given by Rawls (1971), and alternatively by Nozick (1974). These are derivations from the works of the European Enlightenment thinkers, more specifically Kant (1977) Generations of enlightened thinkers have proceeded therefrom (Starr, 1984). Starr wrote along the juristic lines, a topic that is inextricable with near to fiqh of shari'ah origin:

Rawls defined social justice as follows (Stanford Encyclopedia of Philosophy, 2008). Rawls' theory of *justice as fairness* describes a society of free citizens holding equal basic rights and cooperating within an egalitarian economic system. His theory of *political* liberalism delineates the legitimate use of political power in a democracy, and envisions how civic

M. A. Choudhury, *Tawhid and Shari'ah*,
https://doi.org/10.1007/978-3-030-49087-4_4

unity might endure despite the diversity of worldviews that free institutions allow. His writings on *the law of peoples* set out a liberal foreign policy that aims to create a permanently peaceful and tolerant international order.

Unlike Rawls, Nozick defined social justice in terms of the theory of minimal state in the following way (Stanford Encyclopedia of Philosophy, 2018) as follows: "Nozick argued that respect for individual rights is the key standard for assessing state action and, hence, that the only legitimate state is a minimal state that restricts its activities to the protection of the rights of life, liberty, property, and contract."

Such exactness in the definitions of terms of moral values has given rise to subtle academic studies and legal meanings, as is the case of writings discerning the difference between the meanings of social justice and fairness in Rawls (Wolff, 1977), and the neoclassical roots of Nozick's (2003) theory of justice.

Likewise, on the theme of equity there has for long now existed the idea of wage determination and the excess value of capitalist prices on goods and services above the marginal wage theory of labour (Cole, 1966; Mandel, 1971). Based on either the firm's level of such an excess acquisition by the capitalist producer or the macroeconomic idea of aggregate composite wage bill, the application of the corresponding meanings of equity would be taken up in the study of law and economics (Posner, 2007).

In conclusion to this introductory section, we note that Islamic juristic (fiqh) and its relationship with Islamic law, the shari'ah, convey humanly felt meanings to terms of moral values. Such rationalistic derivations of moral meanings leave them not necessary in consonance with the Qur'anic meanings. Thereby, the relevance of the Qur'anic origin of moral meaning, found to be imminent in the primal ontological law of monotheism as a dynamic universal law of Tawhid (sunnat Allah) governing everything in generality and particulars (a'lameen), is ignored completely. Consequently, the ontological meanings of Qur'anic terms, such as 'justice', 'wellbeing' (maslaha), 'public preference' (istihsan), 'opinion' (ra'y and fatawa), and 'value' (masalih) have been relegated to human concoction. The shari'ah has been distanced from the otherwise indispensable ontological origin of Tawhid as the dynamic law in conjunction with the generality and particularities of the world-system. This state of delink between the humanly concocted contents of shari'ah and the Qur'anic origin of Tawhid as law has deepened in Islamic legal history of the shari'ah (Coulson, 1984).

The resulting dichotomy between the Tawhidi primal ontology as a priori domain of reasoning and the construction of the Tawhidi-induced law of unity of knowledge of the world-system as a posteriori domain of morally embedded materiality of world-system has marred the essential possibility of shari'ah in Tawhid. This result is similar to the philosophy of rationalism qua Kantian heteronomy (Wikipedia, visited 21 Aug. 2018. https://en.wikipedia.org/wiki/Heteronomy). It is denoted by [a priori = Qur'an and sunnah in respect of unity of monotheistic law, Tawhid] ∪ ∩ [a posteriori = world – system] = ϕ. To enforce the moral will on human choice is the problem of heteronomy. Heteronomy is contra rationalism. To deny heteronomy in respect of the endogenous embedding and interactive and integrative relationship between a priori and a posteriori reasoning is the foundation of rationalism as the human-concocted worldview.

The inference to be drawn from this result on heteronomy for shari'ah is that, like Kant on the Categorical Imperative of the moral law, shari'ah parri passu is premised on the Qur'an, and thus ought to refer to Tawhid as law; yet it has not been able to derive and sustain the dynamics of the Tawhidi law in its immersion in the morally induced world-system by the force of monotheistic unity of knowledge as the primal ontological beginning. The opinions and juristic rulings of the interpreters and Islamic schools of thought (mujtahids) became the rule of the shari'ah. Thereby, Tawhid as law emanating from the Qur'anic primal ontology of unity of knowledge (Qur'an, 36:36), though not intended, was diluted and forgotten (Ghazali, trans. Karim, n.d.). Terminology of the mujtahids, like maqasid al-shari'ah and shari'ah-compliance in recent times became the hidden mainstay of rationalism in shari'ah. Thereby, as well as the various moral values indentified with the juristic and shari'ah law, as mentioned earlier, all fell into imprecision of ra'y and human-designed meanings of istihsan and maslaha, rather than being derived continuously by ijtihad based on Qur'an and sunnah with the discourse of learned participants but as an optional medium of deriving rules and meanings.

Objective

The objective of this chapter is to revert the true premise of Islamic methodological worldview to Tawhid as the primal ontological law and shari'ah as the codification of practices and choices under Tawhid. This to us will mean the secondary rules of the primal ontological law of Tawhid vis-à-vis the Qur'an and sunnah.[1] The Tawhid methodological worldview will be shown to present the full model of Tawhid dynamics as law and its formalism and application to the details of the world-system. An example from Islamic economics is examined to bring out the contrariness of solely starting deductive reasoning from shari'ah as opposed to Tawhid as law in terms of its ontological, epistemological, phenomenological, and sustainability dynamics in relationship to the general comprehension of the world-system. Thereby, the precise identification of the moral terms, as of justice, fairness, and equity and other ones are taken up within the maslaha (wellbeing) study.

Review of the Literature: Problem of Specification of the Islamic Methodology by shari'ah

Problems of specification of Islamic worldview by the foundation of Tawhid as law are found in Kamali's characterization of the relationship between shari'ah, qiyas as analogy, istihsan as public preference, ra'y as opinion, and maslaha as wellbeing function that lends itself to juristic (fiqhi) and usuli (of usul al-fiqh) kinds of development of ijtihad. Kamali (2005, p. 75) writes concerning such a problematic relationship that refers primarily to the variant interpretive approach of the mujtahdis and the imams of the various Islamic schools of thought: "But since qiyas is founded on a particular effective cause, which is often identified by the mujtahid, it is possible that the mujtahid could fall into error, and this is why qiyas is classified as speculative (zanni) evidence."

This observation by Kamali on the history of shari'ah points out that even though the origin of shari'ah is claimed to be in the Qur'an and sunnah, which would thereby mean invoking the primal ontology of Tawhid I (Ghazali, trans. Karim, n.d.), yet ever since its early intent shari'ah became splintered and speculative. The rise of Islamic sects, their mujtahids, and the prevalence of kings and their dynasties, as of the Abbaside and Fatimide rulers, followed by latter days' Wahabi and Ahlal sunnah movements, and subsequently the influence of regional customs (urf and ada) on the emergence of shari'ah (Hooker, 1984) have all contributed to the opinion and preference oriented ijtihad in shari'ah. As a result, the complementary practices of istihsan, qiyas, and maslaha became subjected

[1] Hart, H.L.A. (1994). *The Concept of Law*, p. 92. Clarendon Press, Oxford, UK.

to similar debilitating effects. Thus, from the very beginning recording back to the eighth century following the demise of Prophet Muhammad (PBUH), the true ontological origin of shari'ah in Tawhid as law disappeared. Never ever was Tawhid accepted as the ultimate universal law of 'everything'. The dynamic methodological worldview of Tawhid as the law of unity of monotheistic knowledge was never searched and discovered. Over the years shari'ah has failed to link in continuity with its earlier intended basis of Tawhid, that is the Qur'an and sunnah. Instead, shari'ah sank by its human-concocted nature. Likewise, all that such a divergent and speculative nature of shari'ah touched subdued the affected disciplines into madhabi thought. There were differences between the Maliki and later on Shafei schools that rejected the validity of istihsan, whereas the Hanafi school of fiqh permitted istihsan. A pronounced example of such a case is the so-called field of Islamic economics and finance. The matter of Islamic financial instrument of murabaha (mark-up pricing) is an example of debate over its nature of riba-like financing (JOSR-IEF).[2] Also see *IOSR-Journal of Economics and Finance* (IOSR-JEF).

In present days' Islamic reflections, the reference to Tawhid in shari'ah is totally absent. This is to be noted in the works of Ashur (2013) and Attia (2008).

Despite that Ashur has written well on the higher objective of maqāsidal-sharīᶜah, and in the treatment of the opposite to maṣlaḥa, namely mafasid, to which ribā belongs, the focus of Ashur's study remains on the legislative (fiqhi) aspect of Islamic law in regards to human transactions. The vaster inter-causal relations between the divine law as Tawhid and the philosophy of balance, and the study of interactively generated effects of trade against ribā affecting maṣlaḥa, remain implied. Such studies are constricted merely to descriptive text. The most important issue of explaining the nature and abolition of ribā in relation to the whole gamut of Islamic variables in the form of inter-causality between maṣlaḥa variables and the formal triangulation, as opposed to sheer statement of verses, sunnah, and references from the Islamic scholars, has not seen the exegetic orientation of the studies on trade versus ribā. Yet such a relational study of abolition of ribā and the explanation regarding the principle of complementarities between the good things of life, such as fairness, justice, and trade,

[2] Baktiar, A. & Aedy, H. (2017). "Murabaha implementation in Islamic Bank (Study at Bank Muamalat Kendari Branch)", IOSR Journal of Economics and Finance, 8:5, 2017, PP 13–27.

contrary to the relational consequence of disappearance of ribā occurring automatically in the face of trade, has not been studied.

Attia's erudite study of maqāsidal-sharīᶜah, well-intended though to explain a system study of Islamic law, yet loses the fundamental groundwork of treating ribā and the Islamic participatory financial economic instruments by means of an organic system of contrast proceeding by cause and effect. The result thereby is a failure to recognize the importance of systemic formalism of such a vast nexus of relations for modelling policy-theoretic and institutional effects of a ribā-free economic and social system, and to explain thereby, the scientific implications of such a vast nexus of inter-causal relations. Such implications of the study of usury in the Qurʾān are essential for formal study of the nature and prohibition of ribā and the permissibility of trade. They are derived by Qurʾānic exegesis in the extended triangulated ways of studying the inner dynamics of trade versus riba.

The inference of this section points out that the unsettled human-concocted nature of shari'ah in all its forms and of the juristic approach of fiqh, fatawa, ra'y, maslaha, and istihsan could not live up to the mujtahids' declaration that, contrary to its claimed roots in the Qur'an and sunnah, the evolution of shari'ah as the Muslim claim to be Islamic law, ceased to maintain any and every trace of the Qur'anic worldview premised on Tawhid as the true ontological law of monotheism and its distinctive phenomenology concerning the details of the world-system. Hallaq (2005, p. 3, 2009) points out the resulting transformation of shari'ah over time: "In the case of Islamic law the essential attributes—those that gave it its shape—were four: (1) the evolution of a complex judiciary, with a full-fledged court system and law of evidence and procedure; (2) the full elaboration of a positive legal doctrine; (3) the full emergence of a science of legal methodology and interpretation which reflected, among other things, a large measure of hermeneutical, intellectual and juristic self-consciousness; and (4) the full emergence of the doctrinal legal schools, a cardinal development that in turn presupposed the emergence of various systemic, juristic, educational andpractice-based elements."

Laudable though these developments are to bring shari'ah into the light of its modern treatment, yet the trend moved it in unwanted ways towards secularization away from the Tawhidi dynamics of this ontological law. The premise of Qur'an and sunnah could not be ever upheld in the light of Tawhid in respect of its dynamic methodological worldview. All that was left of a secular orientation of the Qur'an were reliance on certain verses, which the mujtahids capitalized to develop their scholastic exegesis, ray, qiyas, and istihsan to match up with other imminent elements. One such important element was maslaha as understood from the fiqhi and fatawa viewpoints. The most irreconcilable cleavages were of the Muslim rationalist school of kalam and the theocratic school of mutakallimun (Qadri, 1990), and of the Sunni and Shi's in respect of differencing in the Qur'anic worldview of Tawhid. This otherwise ought to have been unified worldview (Al-Saduk, 2009; Choudhury, 2014a; Wahab, 2012). The recent work on Tawhid by Professor Ismail Faruqi (1992) could not manifest the Tawhidi dynamic methodological worldview of unity of knowledge and the world-system. The universal comprehensive nature of Tawhid as the ontological law of Qur'anic 'everything' could not be allowed to overarch the exclusively juristic (fiqhi) nature of shari'ah.

Thereby, the Qur'an and sunnah taken together could not stand up as the methodological praxis of the Islamic worldview. Shari'ah with its components in usul al-fiqh and the subsequent development of fiqh as practicing codes of shari'ah as law only referred to the exegeses of the Qur'an, but not to its ultimate universal and unique essence. This singular message of the Qur'an in its strictest monotheistic message is Tawhid, which shari'ah could not discover in its dynamic and unique and universal nature in the order of 'everything'. Centuries have passed in this misguided venture, causing cleavages and interpretive problems in shari'ah, contrary to the Tawhidi methodological worldview.

In regards to maqasid al-shari'ah upon which the field of shari'ah has developed, the focus has traditionally remained narrowed down on worldly affairs called muamalat only. All of the big questions, such as the relationship of cosmology and the consideration of the meta-science of primal ontology leading to epistemology, phenomenology, and continuity of the analytics of Tawhidi epistemic have not drawn the concern of the traditionalists—even within the domain of muamalat. Likewise, the great watershed of technical formalism, analysis, and other sophistications has remained unattended. Examples are the case of information technology, specification of the wellbeing function called the maslaha studied in the midst of system complexity to answer complex problems of interaction between the cosmological order and the earth in respect of ethical and moral issues. Only recently, attention to this systemic study of maqasid al-shari'ah is being given by Auda (2008) and Choudhury (2017, 2018). See also Hawley (1986) on the possible human ecology design of Tawhidi systemic model, which is not to be found in the study of shari'ah in any of its forms.

The final result of the above-mentioned kind of developments in shari'ah contra search and discovery of the dynamic methodological worldview from Tawhid as the law (Choudhury, 2017) is to bring us to the point of deriving the Tawhidi approach yielding the precise definitions of moral values. We now turn to this methodological inquiry concerning Tawhid methodological worldview contra shari'ah and its fiqhi composition.

Methodological Inquiry

Flowing out of the above-mentioned arguments, this section formalizes the methodological explanation of Tawhid as the primal ontological super-encompassing shari'ah as a theocratic and juristic development. This state of affairs in the evolution and repetition of the nature of shari'ah cannot cause the shari'ah to be law. Rather, according to the theory of law (Hart, 1994), it constitutes a field of codification of the ahkam (rules) that arise from the formal methodological worldview of Tawhid as the primal ontological law of 'everything'.

Here is a logical basis of rejecting the long-standing belief on shari'ah to be the primal Islamic law. Firstly, shari'ah is claimed to be derived from the Qur'an. In that case, Tawhid as primal law arising from the Qur'an, and mapped on to the experiential world-system, then must comprehend the primal law and determine the be-all and end-all of Shari'ah.[3] Then those parts of shari'ah that form usul al-fiqh as human-concocted madhabi qiyas and juristic preferences istihsan, ra'y, and maslaha cannot be justified by the rationalistic invlinations, and do not arise from the Qur'anic law of Tawhid and the world-system.

Thus, in all, the limitation of shari'ah to the following debilities makes this legalistic and theocratic approach untenable as the law. At best shaari'ah remains a codification field of choices of Tawhidi possibilities: (1) Tawhid, contrary to shari'ah, comprehends the entirety of 'everything'

[3] Kamali, H. (2003, p. 3): "An adequate grasp of the methodology and rules of interpretation also ensures the proper use of human reasoning in a system of law which originates in divine revelation." According to the Qur'an, nothing of experience is left out of the Qur'an (kitab al-mubin). In that case, Tawhid, as it comprehends the ultimate universal and unique message of the Qur'an, also encompasses 'everything' in a substantively explained way.

between the heavens and earth and all that is between, and below the earth, and above the earth, considering only muamalat and not the great relationship of muamalat with the cosmological world-system (Qur'an)[4]; (2) Shari'ah faces its inability to derive a unique and universal methodology and moral material embedded method from the knowledge of Tawhid as the primal ontological law of the Qur'an and its transmission by the sunnah; (3) The variegated nature of usul al-fiqh in shari'ah by differentiation of madhabs and historical developments, and all that this humanly concocted field bestows remains contrary to the eternity of Tawhidi law of 'everything'; (4) Being unable to define the dynamics and formal worldview of the methodology of unity of knowledge and its analytical method derived from the Tawhidi ontological law, shari'ah has left unanswered several critical questions of the Qur'an.

Examples of such a case are the absence of definition and method of measurement of interest rate (riba) for avoidance in the intertemporal case of asset valuation in the light of Tawhidi law of monotheistic unity of knowledge. Consequently, the formalism of the maslaha as a functional delineation of the Tawhidi methodology is ignored for preference of juristic and theocratic constriction of the terms of istihsan, qiyas, ra'y, and masalih. On the other hand, it is accepted by taqlid (uncritical acceptance of authority) that shari'ah embodies the legal completeness of the Qur'an and out of this into juristic extension. If truly shari'ah embodies the whole of Qur'anic law in this sense, then it must equate ideally with Tawhid as law. Some of this law would be methodologically explicit; the remainder would be derived but yet from the self-same methodology of unity of knowledge and the totality of world-system according to Tawhid as the primal ontological law. The question then is critically this: Why would the true and originary Islamic law be termed and accepted as Tawhid instead of shari'ah? Indeed, the latter has no ontological, epistemological, and phenomenological totality of methodology as has Tawhid in respect of its monotheistic methodology and worldview of unity of knowledge. Expressions below bring out the contradiction that arises by holding up shari'ah as law against the ultimate universality and uniqueness of Tawhid as the ontological law.

[4] Qur'an (45:36): Then Praise be to Allah, Lord of the heavens and Lord of the earth—Lord and Cherisher of all the Worlds!

The Shari'ah Orientation Relating to Tawhid (Qur'an and Sunnah)

Usul-al fiqh and its derivatives start here.

[Tawhid	$\rightarrow_S$	$\rightarrow \{\theta,\mathbf{x}(\theta)\}]$: ↓shari'ah↓ ⇔ all of Qur'an and sunnah ⇔	Tawhid
(Qur'an & Sunnah)	Qur'anic	world-system as understood	⇓
Ontology, epistemology	transmission	along with underlying usul	Qur'anic contradiction
and phenomenology of	by means of	al-fiqh and every idea in	of shari'ah~ Tawhid
unity of knowledge	sunnah	every other derived concepts	
knowledge			

(4.1)

The Orientation of Tawhid as Law Contra Shari'ah[5]

1	2 3	4	5	6
[Tawhid$\rightarrow_S$]: $(\Omega,S)\rightarrow_f$	$\theta^*\rightarrow\{\theta,\mathbf{x}(\theta)\}$:	$\rightarrow_W W(\theta,\mathbf{x}(\theta))$	construction[42]	Process
(Qur'an & Sunnah)	Qur'anic	world-system in	and evaluation	ends and a
Ontology, epistemology	knowledge	generality and	by maslaha as	new process
and phenomenology of	by sunnah	specifics studied	wellbeing function	commences
unity of knowledge	and discourse	under (Ω,S);	$W(\theta,\mathbf{x}(\theta))$ In the light	in continuity"
		with $\theta\leftarrow\theta^*$.	of (Ω,S), and subject	across
		(Tawhidi law of	to inter-variable	continuums:
		pervasive monotheistic	endogenous	Sustainability
		Unity of knowledge as	circular causation	
		Unraveled by $[(\Omega,S);\theta^*,\theta]$:	system of	7 ⇓
		Properties of evolutionary	relations	evolutionary
		learning induced of the		learning by

(4.2)

Stage 1 and Stage 7 are identical in respect of Tawhid as law of unity of knowledge. Tawhid as law in relation to the world-system in unity of knowledge is explained by pervasive of organic evolutionary learning in the principle of unification between the good things of life, as noted in the Qur'an and sunnah and thereby discoursed by codification in respect of $[(\Omega, S);\theta*,\theta]$, to generate a world-system 'as it ought to be' out of its

[5] The formalism and evaluation method of W(,) is a highly technical approach of estimating and simulating the wellbeing function subject to the endogenous properties of the circular causality between the organically interrelating variables, which explicate the Tawhidi methodology of monotheistic unity of knowledge signified by the pervasive complementarities between the selected variables by the Qur'an and sunnah in $[(\Omega, S);(\theta*,\theta)]$. See for much greater details the many works by Masudul Alam Choudhury.

existing nature ('as is'). This involves the phenomenological method of transition from the system of estimated inter-variable organic relations into a system of unified (participatory, complementary) simulated world-system reconstruction.

There is no relevance of stating the role of intervening shari'ah between Stages 3 and 4, as otherwise is the nature of shar'i ah orientation that does not have a methodology to commence the Islamic law in Tawhid as law.

Between Stages 1 and 7 'everything' comprises the world-system. In the case of Tawhid, it is the law of monotheistic unity of knowledge as the methodology that characterizes the entire organically unified evolutionary learning world-system in its generality and details. The shari'ah orientation has no such methodological analytics.

All the methods derived and consonant with Tawhid as law of monotheistic unity of knowledge and as methodology converge in Stages 4 and 5. Thereby, the meaning and significance of rules arising from maslaha and its implications concerning truth and oppositely in the criterion of falsehood (mafasid) are addressed in the phenomenological Stages 4 and 5. Stages 6 onwards completes the properties of interaction (discourse) and consensus (integration, $(\theta *, \theta)$), and starts evolutionary learning across knowledge, space, and time continuums. Such properties are not to be found in shari'ah orientation.

Justice, Fairness, and Equity as Interrelational Circular Causation Variables Out of n-Dimensional Vector

Expression (4.2) points out in reference to Tawhidi methodological context that values according to the the Qur'an are uncountably many, as the Qur'an (31:27) declares: "And if all the trees on earth were pens and the ocean (were ink), with seven oceans behind it to add to its (supply), yet would not the words of Allah be exhausted (in the writing): for Allah is Exalted in Power, full of Wisdom." Therefore, it is only possible to signify a denumerable set of values according to the existing level of comprehension enabled by the Tawhidi precept of unity of knowledge and subject to incessant expansion, by the vector (also matrix, tensor, etc.) $\mathbf{X}(q) = \{x_i\}[q]$, $I = 1, 2, \ldots, n$; with $\theta \in (\Omega, S; \theta^*)$, and $n(\theta) \in N(\theta)$, expansionary denumerable set of numbers with increasing (decreasing) value of $\{\theta\}$-inducing denumerable number system. Likewise, these values can be continuous in nature. In this case, $n(\theta) \in$ continuous neighbourhood of $N(\theta)$. $\theta \in (\Omega, S; \theta^*)$ have their continuity property. Consequently, $\mathbf{X}(\theta)$ is expandable by analytic functions in the continuously elastic denumerable domain, as by Taylor's series in multivariates. Thereby, the wellbeing function assumes its aggregation in the knowledge, space, and time domains.[6]

In the context of the continuity property in the Tawhidi law of unity of knowledge, the values of equity $(x_1(\theta)\}$, fairness $(x_2(\theta))$, and justice $(x_3(\theta))$ take up their precise definitions within the wellbeing function in respect of the circular causation endogenous relations between these values expressed as mathematical variables in the vector, say, $\mathbf{x}(\theta) = \{x_1, x_2, x_3\}[\theta]$. The specific definitions of these variables are derived from their circular causation relations:

$$x_1 = f_1(x_2, x_3, \theta); x_2 = f_1(x_1, x_3, \theta); x_3 = f_1(x_1, x_2, \theta); \theta = F(x_1, x_2, x_3)[\theta], \quad (4.3)$$

where θ in this particular circular causation equation of wellbeing is the quantitative measure of organic unity of knowledge in the conceptual form of the wellbeing (maslaha) function, $W(\theta) = W(x_1, x_2, x_3, \theta)$. The property of mathematical continuity and differentiability of $W(\theta)$ in terms of the endogenous variables, $\{x_1, x_2, x_3, \theta\}[\theta]$, yields the existence of the implicit function theorem with non-vanishing Jacobian matrix (Hogg & Craig, 1965).[7] This inversion of the wellbeing function now takes the quantitative form as a similar function, $\theta = F(x_1, x_2, x_3, \theta)[\theta]$. The meaning here is that wellbeing is established in degrees by the monotonic levels of complementarities between $x_1(\theta)$, $x_2(\theta)$, $x_3(\theta)$, as evolutionary learning by $\{\theta\}$ progresses on.

Within this comprehensive functional relationship of wellbeing function in respect of the episteme of unity of knowledge, equity is determined in its degrees of θ – induced complementarities with justice and fairness as value attributes. Likewise, it is true with $\{x_2, x_3\}[\theta]$. The inter-variable endogenous organic relational specifications give the moral definitions of the three θ – induced variables.

[6] $E(f(\mathbf{x}(\theta))) = f(E(\mathbf{x}(\theta))) + [(\mathbf{x}(\theta) - E(\mathbf{x}(\theta)))] * f(\mathbf{x}(\theta)) + [(\mathbf{x}(\theta) - E(\mathbf{x}(\theta)))]^2 * f(\mathbf{x}(\theta)) + \ldots + [(\mathbf{x}(\theta) - E(\mathbf{x}(\theta)))]^n * f^n(\mathbf{x}(\theta)) + \ldots \infty\ \theta \in (\Omega, S)$.
Thus, Taylor's expansion exists everywhere within $h(\bullet)$ and across $HH(\theta)$.
$W(\theta, \mathbf{x}(\theta), T) = \int_0^T \int_0^{\mathbf{x}(\theta)} \int_0^{\theta} a_t(\theta)^* [W(E(\mathbf{x}(\theta)) + b_t(\theta)^* Var(x(\theta)] d\theta d\mathbf{x}(\theta) dt$,
encompassing the knowledge, space, and time dimensions.
[7] Jacobian Matrix is well explained by Hogg & Craig (see References).

To include the socio-economic materiality component ($\mathbf{y}(\theta)$ – vector conjointly with the moral θ – induced variables, thereby the endogenous organic ensemble of moral-material variables induced by Tawhidi unity of knowledge denoted by $\{x_1(\theta), x_2(\theta), x_3(\theta), \mathbf{y}(\theta)\} = \mathbf{z}(\theta) = \{\mathbf{x}(\theta), \mathbf{y}(\theta)\}$, the number of circular causation organic relations in terms of Tawhidi unity of knowledge are continued, yielding thus the moral-material specified vector of values in $\mathbf{z}(\theta)$. Statistical observations on details of $\mathbf{x}(\theta)$ – vector can be generated by a range of quantitative values that increase in number with θ – induction. Examples of such quantitative values for moral variables are Gini Coefficeints $\{x_1(\theta)\}$; resource share of development fund at the underprivileged level $(x_2(\theta)\}$; and organic combination of complementarities between $\{x_1, x_2\}[\theta]$ for $\{x_3(\theta)\}$. The above-mentioned explanation of endogenous inter-variable relations of the good things of life is inferred from the Qur'an (36:36; 56:9, 56:13).[8]

Tawhid Contra Shari'ah as the Totality of the Qur'anic Law

Shari'ah scholars and the founders of schools of interpretations of the shari'ah (usul al-fiqh), the mujtahids, according to various sects (madhabs), have committed an illogical conclusion in claiming that all of shari'ah finds its beginning in the Qur'an. This kind of claim is contrary to establishing Tawhid as law, as the shari'ah claim that Tawhid, which is every part of Qur'an, also comprises shari'ah as Islamic law. The illogical nature of this claim is yielded by the following observation. Kamali (2003) writes: "In the sense that legal material occupies only a small portion of the bulk of the text, the Qur'an is not a legal or a constitutional document. The Qur'an calls itself huda, or guidance, not a code of law." If this statement is taken at its face value, then much of shari'ah is humanly derived. The bulk of shari'ah for its claim of legal completeness cannot be the Qur'anic law upon which the origin of Tawhidi worldview is ontologically premised. Yet, the true manifestation of Allah's monotheism is revealed in the Qur'anic law as the primal ontology and completeness of

[8] Qur'an (36:36): "Exalted is He who created all pairs—from what the earth grows and from themselves and from that which they do not know."

Qur'an (56:9). "Observe the weight with equity and do not make the balance deficient."

Qur'an (56:13): "Then which of the Blessings of your Lord will you both [jinn and men] deny?"

Tawhid parri passu. It is for the Tawhidi endowed worldview of unity of knowledge is to be derived in its varied analytical, formal, and applied phenomenology of 'everything'. The Qur'an thus mandates on its votaries to teach deen (religion) with wisdom (hikma).

The Qur'anic declaration that Allah is the Lord of the Heavens and the Earth and all this is between and under and above the earth leaves no corners unturned. The cosmological domain is thereby complemented and explained by the methodological worldview of Tawhidi unity of knowledge, which exists as phenomenological totality of physicalism and non-physicalism in totality. The guidance of Tawhid as the law of methodological worldview of unity of knowledge in the Qur'an of 'everything' (a'lameen) is explained in the derived methodology of unity of knowledge. None of such attributes of universality and uniqueness of knowledge endowment by monotheism as manifest law are any part of shari'ah claimed to be 'Islamic' law. Tawhid by its ontologically established manifest law is thereby extensive, complete, and revelatory of the essence of the Qur'an (dhat). The Qur'an declares, Allah does not go to sleep and slumber. That is, Tawhid as the universal and uniquely all-inclusive law is complete and extended, res cogitans and res extensa for 'everything' (Choudhury, 2015). Such an overarching and transdisciplinary with sublime explanatory power between the heavens and the earth requires greatly analytical power. Tawhid offers this intellection out of its singular primal ontological premise of monotheistic law, namely Tawhidi unity of knowledge. Shari'ah has none of this in any primal ontological form. It is thereby randomly rationalistic. Rationalistic ratiocination in originating Qur'anic interpretation without its legitimate roots in the Qur'an and sunnah. The Qur'an (21:18) declares in this regard: "Nay, We hurl the Truth against falsehood, and it knocks out its brain, and behold, falsehood doth perish! Ah! woe be to you for the (false) things ye ascribe (to Us)."

The inference derived from the above facts is that, despite the unanimous agreement among all mujatahids and Muslim sects originate in the Qur'an and sunnah, the problem remains that the ontological origin of Tawhid is not explained as a methodological entirety with phenomenological applications in 'everything' between the heavens and the earth. Besides, human-concocted implications have entered shari'ah, as if to augment the claimed limitation of legal capacity in the Qur'an. Kamali (2003, p. 25) writes that less than only 620 verses out the total number of 6200 verses of the Qur'an relate to matters of law. Thereby, the universality of Tawhid as law has remained absent in shari'ah. Thereby we write that, {shari'ah claimed edicts of Islamic law} ⊃ {Tawhid as the supreme law ofeverything'}, a contradiction to the Qur'an. Note just a few verses: "and We have revealed the Book to you explaining clearly everything (16:89)"; "a guidance for mankind and clear evidence of guidance and discrimination (between wrong)" (2:185); "And (as for) those who strive hard for Us, We will most certainly guide them onto Our ways" (29:69).

The Ontological Derivation of Tawhid as Law of Unity of Knowledge Contra Shari'ah as Claimed Law

The following schematic derivation of Tawhid as ontological law from the Qur'an is from Kamali (2003). Yet we show that Kamali's derivation of shari'ah from the Qur'an is devoid of a systematic phenomenological order of monotheistic unity of knowledge to the world-system and its continuity over knowledge, space, and time multiverse:

The monotheistic quiddity of the Qur'an is premised on the completeness of unity of knowledge. The first verses of Qur'anic (96, 1–5): revelation are the following: "Recite in the name of your Lord who created—created man from a clinging substance. Recite, and your Lord is the most Generous—Who taught by the pen—taught man that which he knew not." These verses are combined with the following one to establish the supercardinal, all-encompassing, absolute, and complete nature of knowledge embodied in Qur'an, which is the Tawhidi monotheism of unity of knowledge: "Nay it is a glorious Qur'an in a guarded tablet (Lauh al-Mahfuz)" (Qur'an 85:21–22).

We give analytical formalism to the entire ontological, phenomenological, evolutionary learning, and continuum methodological worldview we denote (1) by the following supercardinal topological expression of Tawhid. Tawhid is thus the manifest law in terms of its unravelling of the oneness of Allah in the scheme and order of the divine law of unity of knowledge (pervasive complementarities in the good things of life). The monotheistic law (sunnat Allah) is explained by physical and non-physical holism that overarches the world-system (a'lameen). The a'lameen is thus washed by the signs of Allah (ayath Allah). Thus, Tawhid as law is manifestly rendered to study in the world-system by its explanatory power of generality and particularities of creation of the known and the unknown in the full light of absolute reality. Such a supercardinal completeness and configured possibility of completeness is an understanding of the manifest res extensa of Tawhid as a knowable part of Lauh al-Mahfuz.

The symbolic representation of Tawhid as manifest law of unity of knowledge conveying the oneness of Allah at work in the pervasive extant of the knowledge-induced world-system is denoted by (Ω).

The Qur'anic supercardinal message of unity of knowledge and its dawning on the world-system in totality through the Prophet Muhammad is declared in the Qur'anic verse (26: 193–194): "The Trustworthy Spirit has brought it down. Upon your heart, [O Muhammad]—that you may be of the warners.". Furthermore, there is the verse (42:52): "And thus We have revealed to you an inspiration of Our command. You did not know what is the Book or [what is] faith, but We have made it a light by which We guide whom We will of Our servants. And indeed, [O Muhammad], you guide to a straight path."

We symbolize this section of Prophetic transmission of the Qur'an by his teachings, sunnah 'S' as part of the functionality of Ω by its worldly mapping as,

$$\Omega \rightarrow_S \tag{4.4}$$

The meaning of complete phenomenology as consciousness in terms of Qur'anic knowledge in terms of the manifest law of Tawhid governing 'everything' in respect of unity of knowledge can be derived from the verse (36:12): "Lo! We give life to the dead, and We write down what they used to and what they used to leave. And all things We have gathered in a clear Book."

We symbolize ths part of the Qur'anic universe conveying the primal ontology of the Qur'an by the sunnah of Prophet Muhammad on to unity of knowledge in the world-system comprising moral-material and thereby physicalist and non-physicalist spanning by,[9]

[9] By the Qur'an, $\{\theta*\}$ is discursive knowledge of the learned ones (ulul-amr, tabiun, mujtahid}. Yet the Qur'an (4:59) also declares exclusively to (Ω, S), "O ye who believe! Obey Allah, and obey the Messenger, and those charged with authority among you. If ye differ in anything among yourselves, refer it to Allah and His Messenger, if ye do believe in Allah and the Last Day: That is best, and most suitable for final determination." It is to be noted in the light of this verse that $\{f(\theta,\theta*)\}$ comprises constancy or secondary reference to $\{\theta*\}$, that discursive knowledge derivation in respect to understanding the Qur'anic verses in terms of Tawhid law of unity of knowledge devolves into $\{f(\theta,\theta*)\} = \{f(\theta)\}$; $\{\theta\}$ now denotes the discursively derived and advanced knowledge flows of Tawhidi unity of knowledge to that of all earnest participants in and for (Ω, S). The Qur'an (29:69) declares: "And those who strive in Our (cause)—We will certainly guide them to our Paths: For verily Allah is with those who do right."

$$[(\Omega \rightarrow_S):(\Omega,S)] \rightarrow [_{f1}\{\theta *\} \rightarrow_{f2} \{\theta\}] \rightarrow_{f2*f1=f} \{f(\theta,\theta *)\} \quad (4.5)$$

The Qur'an establishes its model of knowledge-induced consciousness in the phenomenological worldview of Tawhidi unity of knowledge in the moral-material infused process of unravelling in the following verse (29:20): Say: "Travel through the earth and see how Allah did originate creation; so will Allah produce a later creation: for Allah has power over all things." The imminent dynamics of evolutionary learning underlying this verse applies analytically in belief to both the reorigination of the world-system in the great event of Akhira after its demise in temporality and through the continuous processes of change in the world-system, $\{\mathbf{x}(\theta)\}$. Indeed, historical consciousness is marked out in this verse by the unravelling of reoriginative evidences. Such are the experiences of evolutionary learning in continuums of creation.

The continued sequences of the primal ontological origin, followed by its continuous recalling now apply to the generality and particularities of the emergent world-system. The symbolic analytical form continues in its functional derivation across primal ontology, epistemology, phenomenology, and sustained continuums.

$$[(\Omega \rightarrow_S):(\Omega,S)] \rightarrow [_{f1}\{\theta *\} \rightarrow_{f2} \{\theta\}] \rightarrow_{f2*f1=f} \{f(\theta,\theta *)\} \rightarrow_g \{g(\theta,x(\theta))\} \quad (4.6)$$

The 'g(.)' functional here denotes multivariate choices representing the knowledge-induced vector, $\mathbf{x}(\theta)$, which delineate the formalism of the specific problem and issue under study.

The world-system according to Tawhid is blessed by monotheistic unity of knowledge in the symbiotic formalism of $\{g(\theta,\mathbf{x}(\theta))\}$. This causality now forms the bestowal function and is marked by the positively increasing function of the good things of life (maslaha, $W(\mathbf{x}(\theta))$, and is negatively changing in terms of the false things of life (mafasid). The Qur'an (10:64) declares: "For them are glad tidings, in the life of the present and in the Hereafter; no change can there be in the words of Allah. This is indeed the supreme felicity." We symbolize the formal continuity of events by the chain relations,

$$\left[\left(\Omega \rightarrow_{S}\right):(\Omega, S)\right] \rightarrow \left[_{f1}\{\theta *\} \rightarrow_{f2} \{\theta\}\right] \rightarrow_{f2*f1=f} \{f(\theta, \theta *)\}$$
$$\rightarrow_{g} \{g(\theta, x(\theta))\} = W(x(\theta)). \tag{4.7}$$

This process is complete at the end of any one intra-system phase of evolutionary learning that is characterized by its specific evaluation procedure, meaning estimation by coefficients denoting elasticity of inter-variable endogenous relations of positive complementarities or negative substitutions. Positive complementarities signify unity of knowledge as participatory inter-variable relations, or between differentiating variables once these are identified by the law of Tawhid.

The maslaha function as wellbeing function is fully versatile in our case contrary to the terminology used in shari'ah to manifest juristic preferences (istihsan). In this case of shari'ah terminology, the Shatibi type basket of social goods results in three kinds of goods that correspond to the five goals of maqasid al-shari'ah.[10] There is no explicit property of inter-relationship between the corresponding variables. On the other hand, the wellbeing (maslaha) function arising from the Tawhidi approach is multivariate in nature in response to the knowledge-induced axiom of unity of knowledge affecting all variables. Consequently, the inherent extendibility of the social goods in respect of the extended goals of maqasid al-shari'ah forms the conceptual and applied nature of evaluation of wellbeing objective criterion function.

Even the concept of juristic preference (istihsan) has changed in Tawhidi methodological model of expression (4.7) from the one given by maqasid al-shari'ah.[11] The endogenously organic interrelations of unity of knowledge between the selected variables causes natural generalization and extension to the systemic variables and the widening goals of diversities

[10] Protection of religion (Tawhid); protection of reasoning faculties; protection of family; protection of progeny; protection of property and property rights.

[11] Groups (i) of similar socio-economic and ethical variables can be grouped together to represent the preferences of agents of such groups, as $\wp_I(\theta)$, $I = 1, 2, \ldots, n$. $\{\theta\}$ are limiting unity of knowledge by ranking of columns of variables in respect of their ranked desirability (importance). These surveyed or ordinal parameters denote group-specific dynamic preferences. The aggregate of such ranked knowledge parameters is represented by $\wp(\theta) = \cup \cap \{\wp_i(\theta)\}$. $\wp(\theta)$ denotes aggregate preference of the group in the form of dynamic integration ($\cap$) derived by systemic interactions ($\cap$) over knowledge, space, and time by evolutionary learning processes; that is, $(d/d\theta)[\cup \cap \{\wp_i(\theta)\}] = (d \wp (\theta)/d\theta) > 0$. Inter-causally unified relations give the meaning of istihsan in Tawhidi model of unity of knowledge.

between the heavens and the earth beyond what maqasid al-shari'ah includes in muamalat only. Consequently, many of the sought for indicators, such as maqasid al-shari'ah index, maslaha index, social inclusiveness index of Islamic financing instruments, and the like are subsumed in the Tawhidi methodological and applied approach of evaluation of degrees of unity of knowledge in the wellbeing function.

Inference Drawn from the Tawhidi Contra shari'ah Meanings of Moral and Material Value

Our foregoing discussions elaborate the fact that there are distinctive differences in the meaning, formalism, and application of important terms extracted from the Qur'an as the foundation of Tawhid as law. This is contrary to the terminologies used in shari'ah. In fact, some critical terms can be identified in the shari'ah that fail to have the high grounds of Tawhidi construction of its methodological worldview of universality and uniqueness. One of these important terms is maslaha. In shari'ah, this terminology stands for benefit with its synonym of blessings conveying rewards and benefits. But in assigning the maslaha meaning shari'ah literature nowhere refers to the primal ontology of Tawhid in terms of its substantive origin of organic unity of knowledge along with all the analytical, formal, and applied details of the understanding of the term 'maslaha' as wellbeing in relation to multivariate extensions in the Tawhidi methodological worldview. Along with the distinctive difference in the concept of maslaha between Tawhidi methodological worldview and shari'ah terminology, there is the similar divide in the meaning of istihsan. The formal way of understanding preferences arising from the maslaha is explained in the footnote. Istihsan just as qiyas and ra'y are based on fatawa of usul al fiqh in shari'ah. Contrarily, in Tawhid as law in its methodological worldview of organic unity of knowledge as derived from the quiddity of the Qur'an, and sunnah derives its meanings from the systemic nature of evolutionary learning in interaction and integration with the ultimate institution of a'lameen, world-system. The human reasoning must firstly listen to the inferences gained from this great institutional generality and particularity of the world-system where the Tawhidi unity of knowledge as the ontology rests.

The Consequences of Such Distinctive Differences: An Example

Extensively many verses of the Qur'an can be referred to in unravelling the primal ontological realm of Tawhid as the complete and principal law representing monotheistic unity of knowledge by pervasive complementarities (participations). An example of distinctly different understanding of the concept of riba (financial interest) and trade can be noted in regards to their intertemporal meanings. The Qur'an (3:130) has indeed placed the meaning of avoidance of riba in terms of the intertemporal context: "O you who have believed, do not consume usury, doubled and multiplied, but fear Allah that you may be successful." The edict of avoidance of riba, and thereby the permissibility of trade, is a permanent one. This verse extends that edict of avoidance to the intertemporal plane, for doubling and multiplying of riba (thereby trade and all the good things of life) take place over the knowledge, space, and time dimensions. This is a theme that acquires substantive explanatory space in the dynamics of Tawhidi methodological inquiry contrary to its static treatment in shari'ah. By this same verse, the simple interest rate can be shown to be equivalent to a series of term-structure of interest rates.[12]

There is a significant difference between the concepts of evaluating riba (trade and assets) discretely across periods of time and to evaluate such variables intertemporally across continuums of continuous time. The former approach attenuates to additive valuation of values and functions done separably over space and time of occurrence of contracts. A utilitarian objective criterion is used. A transformation of such a valuation objective function takes the form of the present valuation, in which case the discount rate is the time value of money. This is equivalent to the rate of interest.

On the contrary, the Tawhidi methodology of objectifying the evaluation of the wellbeing function over knowledge, space, and time continuums across knowledge-induced systems $(\theta, \mathbf{x}(\theta), t(\theta))$ results in the following contrasting results:

Total Intertemporal Valuation, $V(x_t, t) = \sum_{s=i}^{m} \sum_{t=i}^{n} V_{st}(x_{s,t})$, where 's' denotes separable contingencies and systems occurring at specific points of time, '*st*'; $s = 1, 2, \ldots, m$; $t = 1, 2, \ldots, n$. $V_{st}(.)$ denotes time-specific discrete system valuations of variables and functions, such as riba, trade, and cash

[12] Term-structure $\{i_s\}$ of simple interest rate 'i'; 'i' is the geometric average of $\{i_s\}: (1+i) = \sqrt[n]{\prod_{s=1}^{n} (1+i_s)^s}$

flows. The utilitarian meaning of this total valuation formula is made by replacing the Vs by welfare (expected utility) functions. A specific version of the total valuation function is the Present-Valuation formula of future cash flows, x_{st} (interest income). We write this formula as, $V(x_t,t) = \sum_{s=i}^{m} \sum_{t=i}^{n} V_{st}(x_{s,t}) = \left[\sum_{s=i}^{m} \left\{ \sum_{t=i}^{n} x_{s,t} / (1+i_{st})_{st}^{t} \right\} \right]$. Here '$i_{st}$' denotes time value of money in the form of cash flows across discrete time-specific systems.

In the case of intertemporal valuation, integrated and evolved interactively over knowledge, space, and time dimensions, the objective criterion of wellbeing in Tawhidi approach replacing $V(\mathbf{x}_t, t, \theta)[\theta]$, now transformed into $W(\mathbf{x}_t, t, \theta)[\theta]$ = convoluted integral $\prod_{st} W_{st}(\mathbf{x}_t, t, \theta)[\theta]_{st} d\theta dx dt$. The wellbeing evaluation in such a continuum means that the evolutionary wellbeing functions across the continuums are evaluated subject to the entire system of circular causation relations in the endogenous variables. Cash flow relation is included therein. Here is an example of asset valuation in respect of the totality of the valuation composition, namely wellbeing by inter-variable and their circular causal relations across the continuums of knowledge and knowledge-induced space and time dimensions. Unless this perspective of continuum asset evaluation is understood, the problem of riba is not solved. Asset life like in the life cycle of consumption and savings and thereby of cash flows of income and wealth is an integral matter of evolutionary learning by intra and inter systemic interactions (*Money Soft*, 'The value-price continuum', https://moneysoft.com/the-value-price-continuum/). This phenomenon must be taken into account to evaluate over the fair, just, and equitable valuation in respect of the compounded objective criterion of wellbeing defined by its Tawhidi reference to unity of knowledge between inter-variable circular causation relations.

Figure 4.1 is a generalized schema of all types of interactive, integrative, and evolutionary learning events occurring in the intra-inter systemic intertemporality in knowledge, space, and time dimensions. Thereby, a specific case is of asset valuation in such intertemporal intra-inter system evolutionary learning dimensions. The evolutionary learning intra-inter systemic continuous relations apply to the common region between interactive, integrative, and evolutionary sets as shown in Fig. 4.1. We consider the intersection between the domains of justice, fairness, and equity marking the complementary balance between these with the properties of interaction, integration, evolution, and continuity or sustainability (IIEC).

A Generalized Schema of Inter-Causality in Tawhidi String Relations

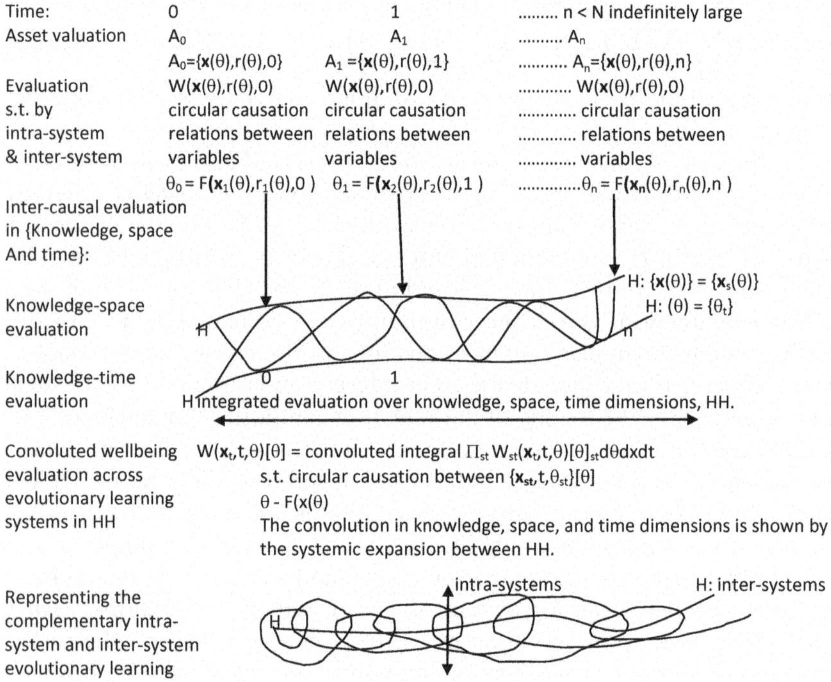

Fig. 4.1 Asset evaluation in terms of evolutionary wellbeing in knowledge, knowledge-induced space, and knowledge-induced time dimensions

The Problem of Determining Interest-Rate in Evolutionary Learning Field in Knowledge, Space, and Time

According to shari'ah, interest rate is avoided in asset valuation in separable intra-system at sequential points in time. An example of this is the enumeration and thereby avoidance of interest rate or the time value of money in progression along separable time periods and thereby intra-systems. Thereby, no attention is given to the evaluation of the wellbeing as a mark of socio-economic value.[13] The result of such a method of asset valuation is to miss sight of the fact that the nature and value of interest rate in the two cases mentioned here are different. In the intra-system calculation of riba, inter-causal relations do not explain the interrelations between the development variables, $\{x(\theta), r(\theta), t(\theta)\}_{st}$ in the evolutionary learning case of intra-systemslinked with inter-systems and $\{\mathbf{x}, \mathrm{r}, \mathrm{t}\}_t$ in the separable case as found in simple terminal valuation formula (Jeans, 1970). Thus, the Tawhidi methodological approach leads to a robust way of assigning value to an endogenous sequence of rates of return in relation to the system of circular causation endogenous variables of the wellbeing function.

The gravity of the differences in asset valuation between the Tawhidi dynamic approach and the shari'ah non-dynamic approach results in several of the financial instruments pronounced by the shari'ah-compliance approach. Examples of such instruments are murabaha and ijara. In both of these cases, the rate set as mark-up or deferred rate of asset payment are exogenously set. The endogenous effect of systemic market action is absent. Because of this problem of incorrectly understanding the rate of interest as rate of return in such financial instruments, all other debt instruments in Islamic financing framework remain contingent in the hidden presence of riba. Takaful (Islamic insurance) is yet another area of asset valuation that is prone to hidden interest rate in the shari'ah context. That is because if premium is to be accumulated for future contingencies of risk, it must be well-determined nearest to its occurrence (Choudhury, 2014b). This being difficult for sustainability of takaful premium fund (tabarru), the contrary way is to set up term insurance with premium funding having the provision of shareholding on unused funds mobilized around primary financing instruments that are not debt instruments. Indeed, it is essential to clean up the Islamic financial portfolio of debt instruments to avoid the hidden presence of riba as explained above. The Qur'an (3:130:1) declares: "O believers, take not doubled and redoubled interest, and fear God so that you may prosper." Such kinds of riba are hidden in the intertemporal evolutionary learning nature of asset valuation of the sequential additive nature of intra-systemic valuation.

Now we return to the problem of identifying the Islamic values, such as of equity, fairness, and justice in asset valuation. We note that, as long as

[13] The example pointed out in the text of this chapter is valuation as, $V(x_t, t) = \sum_{s=i}^{m} \sum_{t=i}^{n} V_{st}(x_{s,t})$, with $V_{st}(x_{s,t}) = A_0(1 + r_{st})^{st}$. No mention of wellbeing as measure of value in knowledge, space, and time dimensions is mentioned. Consequently, no relevance is paid to intra-system and inter-system of evolutionary learning by way of circular causation variables.

the hidden interest rate abounds in the sequential form of asset-valuation over time while ignoring the intertemporal valuation across intra-system and inter-system, the aspiration for the mentioned values cannot be resolved. Besides, like the absence of endogenous inter-causal relations in identifying riba within the development vector there also remains this sustainability nature of asset valuation in shari'ah approach. This is contrary to the approach of Tawhidi methodological worldview. Tawhid as primal ontological law of organic unity of knowledge is thereby contrary in its methodology to the shari'ah viewpoint.

We summarize this section with the following formal explanation of the contrary gap between the Tawhidi ontological worldview of organic unity of knowledge and the piecemeal treatment of the so-called 'Islamic' rules by shari'ah.

Moral values: Equity, fairness, justice

$$\updownarrow \text{Embedding by unity of knowledge}\{\theta\}$$
$$:(\text{Tawhid as Law})\cap(\text{Shari'ah}) = \phi \text{ in this respect}$$

Material values: interest rate in its fullest sense of asset valuation.

Conclusion

The important conclusion to be drawn out of the above discussion is that the claim of shari'ah as Islamic law is misunderstood and misguided. The fact is that other than the common rules of personal and social nature that shari'ah complies with the Qur'an, in which there is no dispute, the greater domain of the heavens and the earth and all that is below the earth and above the earth, as the Qur'an comprehends, shari'ah is totally silent in such interdisciplinary essential matters. Thus, shari'ah unlike Tawhidi methodological worldview in 'everything' interconnecting the heavens and the earth in generality and details, shari'ah has no methodological approach in these great matters. Shari'ah merely adopts certain rules on matters of discourse concerning muamalat. The reference to the cardinal foundation of Tawhid remains merely implied with no methodological interpretations and uses in shari'ah for explaining the activities in the world-system. Thus, the inner sophistication of substantive methodology of Tawhid as law and the socio-scientific methods of transdisciplinary worldview of unity of knowledge is not to be found in shari'ah since its inception around 200 years after the demise of Prophet Muhammad. The analytical method like that of evaluation of maslaha, subject to circular causation relations between the endogenous interrelated variables or its like, remains absent in shari'ah.

Such problems with shari'ah in all its forms have deepened, not resolved, over the entire span of time of its existence. The differences in unresolved issues by the treatment of shari'ah in respect of its madhabi (sects) abound. Examples are of the cantankerous disagreement on issues of istihsan (juristic preference), qiyas, ijtihad, r'ay, and fatawa that remain abound in usul al-fiqh, a truly humanly derived innovations of interpretation and meaning. The ambivalence of Tawhid as law is extensive in shari'ah and the human-concocted discourse that is not even made the reference point in the study of the shari'ah of nations over space and time. In this regard, Kamali (2003, p. 307) writes, yet with utter forgetfulness of the great difference on the methodological understanding of Tawhid between the earlier generations and the ummah of Prophet Muhammad: "This is especially so with regard to the practical rules of shari'ah, that is, the ahkam (rules derived by fiqh and fatawa), in which the shari'ah of Islam is self-contained." Thus, the shari'ah instead of Tawhid is claimed by the generations of Muslim sects (madhabs) to be the law, contrary to the Qur'an and the sunnah in their unequivocal upholding Tawhid as the law of 'everything'. Thus, Kamali (2003 op cit) writes, "The shari'ah in other words, is the exclusive source of all law for the Muslims". This is far from the truth regarding Tawhid contra shari'ah as the ultimate law by the primal ontology of monotheism in Tawhid as law. The similarity with the previous generations abiding with monotheism is in regards to Tawhid.[14] However, the worldview of methodology and derived methods of application remained different between Islamic and the other religions. This divergence made the irreconcilable difference in historical consciousness regarding the ultimate law governing nature, scheme, and purpose of 'everything'. It is therefore illogical why shari'ah and not the Qur'anic meaning of Tawhid as the phenomenological law must reign over the ultimate legal framework of knowledge in Islam. It is the misunderstanding on this ground that has caused almost complete oblivion of Tawhid from

[14] Qur'an (5:48): "For each We have appointed a divine law and a traced-out way. Had God willed, He could have made you one community. But that He may try you by that which He has given you. So vie one with another in good works. Unto God you will all return, and He will then inform you of that wherein you differ."

the phenomenological construct of the Islamic worldview. Tawhid has thus been limited to certain belief rather than manifestations of actions in the totality of the world-system. Indeed, Tawhid as the inscrutable law of 'everything' is at once the transdisciplinary methodological ontology of science and society as of personal needs. Science stands for guidance as anything else—by the Qur'an and sunnah. This principle of generality and its details is totally absent in principle and practice in the relationship of shari'ah and the world-system. On such a principle of social and scientific nature of reality writes Inglott (1990) on Whitehead's idea: "Whitehead's philosophical understanding of the universe as an interconnected web of relations, as well as the ontological nature of the relational self offer a new paradigm of human society. In contrast to the individualism of the liberal tradition, process philosophy defines human society as a relational structure of experience. ... "The present holds within itself the complete sum of existence, backwards and forwards".

Because of the absence of precise definitions and measurements, shari'ah is unable to attain such goals for the many attributes of goodness and blessings conveyed by the Qur'an for the good things of life (halal al-tayyabah). Contrarily, Tawhid is able to accomplish such precise explanations by virtue of the inter-causality between the endogenous inter-variables. Thus, equity, fairness, and justice are individually estimated and reconstructed as values in terms of the rest of the variables of choice. On the other hand, each of the other variables is estimated and constructed in terms of their relationships with the specific values of equity, fairness, and justice. The emergent complementary relations between all the variables together explain the definition by balance of inter-variable causality.

We take here the example of the vector, $\{R, E, F, J\}[\theta]$, where $R(\theta)$ denotes knowledge-induced resource value, E denotes equity variable, F denotes fairness variable, J denotes justice variable. Each of these variables and all other ones in the enlarged vector of variables are induced by 'θ'. Thereby, evolutionary learning induces the expendability of the variables. In the log-linear form of the circular causal relations, equity is defined by the quantitative measure and partial elasticity coefficients $\in_{ij}$ of ith variable with respect to jth variables, such as, $[(\partial R/R)/\partial E/E]$; $[(\partial E/E)/\partial R/R]$; $[(\partial R/R)/\partial F/F]$, $[(\partial F/F)/\partial R/R]$; $[(\partial R/R)/\partial J/J]$; $[(\partial J/J)/\partial J/J]$; and between all the other variables. The pairs do not equate to inverse elasticity values.

These precise elasticity coefficient values are obtained by the evaluation of the wellbeing function (maslaha) subject to circular causation relations between the variables. Such a evaluation function was given earlier. The estimated (simulated) elasticity coefficients convey the exact measures of percentage change in resources in respect of percentage changes in equity as fair distribution of equitable resources; of the same for equitably just treatment of every individual per unit of percentage distribution of resources; of socially just distribution of percentage changes of resources; and of all other variables in percentage changes to each other within the total inter-variable complex of interrelations. No such approach to precision is available in the edicts of shari'ah except for open definitions. A work on fairness and justice on qualitative explanation and quantitative measurement is of Rawls (1971, op cit).

References

Al-Saduk, A.-S. (2009). Tawhid in Theological Mode. 35–49 (by Morrow, J.A. & Vittor, L.A.). In *Kitab al-Tawhid* (The book of Divine Unity) (Eds.), Morrow, J.A.R. Rizvi. Parastesh, The Savior Foundation.

Ashur, M. T. I. (2013). *Treatise on Maqasid as-Shari'ah*. Herndon, VA: International Institute of Islamic Thought.

Attia, G. E. (2008). *Towards Realization of the Higher Intents of Islamic Law: A Functional Approach of Maqasid as-Shari'ah*. Herndon, VA: International Institute of Islamic Thought.

Auda, J. (2008). *Maqasid as Shari'ah as Philosophy of Islamic Law* (p. 45). Herndon, VA: The International Institute of Islamic Thought.

Choudhury, M. A. (2014a). *The Socio-CyberneticStudy of God and the World-System*. Global, PA: Ideas Group Inc.

Choudhury, M. A. (2014b). *Tawhidi Epistemology and its Applications: Economics, Finance, Science, and Society*. Cambridge: Cambridge Scholars Publishing.

Choudhury, M. A. (2015). Res extensa et res cogitans de maqasid as-shari'ah. *International Journal of Law and Management, 57*(6), 662–693.

Choudhury, M. A. (2017). *Absolute Reality in the Qur'an*. New York: New Palgrave.

Choudhury, M. A. (2018). *A Phenomenological Theory of Islamic Economics*. Kuala Lumpur: University of Malaya Press.

Cole, G. D. H. (1966). *The Meaning of Marxism*. Ann Arbor, MI: The University of Michigan Press.

Coulson, N. J. (1984). *Commercial Law in the Gulf States: The Islamic Legal Tradition*. London: Graham & Trotman.

Faruqi, I. (1992). *Al-Tawhid: Its Implications for Thought and Life*. Herndon, VA: International Institute of Islamic Thought.

Ghazali, I. (n.d.). *Ihya Ulum-Id-Din* (F. Karim, Trans. & Eds.). Vol. 5. Lahore: Shah Muhammad Ashraf.

Hallaq, W. (2005). *The Origins and Evolution of Islamic Law*. Cambridge: Cambridge University Press.

Hallaq, W. (2009). *Shari'a*. Cambridge: Cambridge University Press.

Hart, H. L. A. (1994). *The Concept of Law* (p. 92). Oxford: Clarendon Press.
Hawley, A. H. (1986). *Human Ecology*. Chicago, IL: The University of Chicago Press.
Hogg, R. V., & Craig, A. T. (1965). *Introduction to Mathematical Statistics*. New York, NY: Macmillan Co.
Hooker, M. B. (1984). *Islamic Law in South-East Asia*. Oxford: Oxford University Press.
Inglott, P. S. (1990). The Rights of Future Generations: Some Socio-Philosophical Considerations. In S. Busuttil, E. Agius, P. S. Inglott, & T. MacelliEds (Eds.), *Our Responsibilities Towards Future Generations* (pp. 17–27). Malta: Foundation for International Studies & UNESCO.
Jeans, W. H. (1970). Borrowing and Capital Structure. Chap. 6 in *The Analytical Theory of Finance*. New York: Holt, Rinehart and Winston, Inc.
Kamali, M. H. (2003). *Principles of Islamic Jurisprudence*. Cambridge: Islamic Text Society.
Kamali, M. H. (2005). *Equity and Fairness in Islam*. Cambridge: The Islamic Text Society.
Kant, I. (1977). In C. J. Friedrich (Ed.), *The Philosophy of Kant: Immanuel Kant's Moral and Political Writings*. New York, NY: The Modern Library.
Mandel, E. (1971). *The Formation of the Economic Thought of Karl Marx, 1843 to Capital* (B. Pearce, Trans.). New York, NY: Monthly Review Press.
Nozick, R. (1974). Distributive Justice. *Philosophy and Public Affairs, 3.*
Nozick, R. (2003). *Invariances, the Structure of the Objective World*. Boston, MA: Belknap Press of Harvard University Press.
Posner, R. (2007). *Economic Analysis of Law*. New York, NY: Aspen Publications.
Qadri, C. A. (1990). *Philosophy and Science in the Islamic World*. London, Eng: Routledge.
Rawls, J. (1971). *A Theory of Justice*. Cambridge, MA: Harvard University Press.
Stanford Encyclopedia of Philosophy. (2008). John Rawls. Retrieved September 15, 2018, from https://plato.stanford.edu/entries/rawls/
Stanford Encyclopedia of Philosophy. (2018). Robert Nozick's Political Philosophy. Retrieved September 15, 2018, from https://plato.stanford.edu/entries/nozick-political//
Starr, W. C. (1984). "Law and Morality", in H.L.A. Hart. *Legal Philosophy, Marquette Law Review, 67*(4), 613–689.
Wahab, S. A. (2012). *Kitab At-Tauhid, the Book of Monotheism*. Jeddah: Darussalam Publishers & Distributers.
Wolff, R. P. (1977). *A Reconstruction of a Critique of a Theory of Justice: Understanding Rawls*. Princeton, NJ: Princeton University Press.

CHAPTER 5

Tawhid as Law Contra Shari'ah in Standardizing a Theory of Islamic Financial Economy

Introduction

This is a short chapter focusing on the problems encountered by 'shari'ah-compliance' idea and practices of Islamic banking on the issues of standardization of rules and opinions concerning unification of financial fatawa. The chapter presents the authentic alternative means that could be adopted by Islamic banks to universally harmonize 'shari'ah-compliance' fatawa and practices in Islamic banking. These areas are now riddled with problems such as the following one: prevalence of the problem of interpretation of jurisprudence (fiqh) by diversity of views of Islamic schools. This is particularly noted for the Shafei school (madhab) in Malaysia, and oppositely by Hanbali and Hanafi madhabs in the Arab region and elsewhere. A concern of this chapter is on the topic of standardizing of fiqi interpretations and practices of Islamic banks. Along with this is the question: How can Islamic banks adjust their financial fatawa within the international monetary and financial system? How can fatawa be applied to matters of financial loans, deposits, and commercial matters including market exchange issues? How can multiplicity of fatawa, sometime by the same mujtahid (e.g. Shafei, Shatibi) and their contending juristic consequences on 'shari'ah-compliance' and product development be treated for unification under a standarized shar'iah rule? In answers to all of these questions, the chapter seeks for alternative views on financial pricing.

M. A. Choudhury, *Tawhid and Shari'ah*,
https://doi.org/10.1007/978-3-030-49087-4_5

Objective

This chapter is simply a critique of the approach to standardization of fatawa around madhabi diversity and the alternative recommendation regarding practices of muamalat in its broad sense of finance, banking, and market exchange. The criticism here points out that the idea of 'shari'ah-compliance' cannot be upheld in the light of the epistemic worldview of what truly is Islam when unequivocally understood and furthered upon her most indelible law that applies to 'everything'. This is the law of Tawhid, oneness of Allah as the episteme of unity of knowledge as methodology leading to applications and inferences.

By Tawhid as law, this chapter means the primal ontology of monotheistic unity of knowledge, and its induction and inquiry of all details of the world-system. Such world-systems are muamalat, money, finance, and Islamic banking. We will examine how Tawhid as law can be uniquely and uniformly studied in and for standardization. Tawhid as law has a distinctive methodology that applies to formalism, interpretation, application, and sustainability of unity of knowledge in the framework of evolutionary learning on monotheistic unity of knowledge. While the vast details of the field of Tawhidi methodological development as primal ontological law and applications are not covered in this chapter, the topic concerning these can be found in the works of Choudhury. Some of these are Choudhury (2006, 2014, 2016a, 2016b, 2017).

This chapter only picks up a few points of contention on the way towards financial standardization of juristic interpretations. By Tawhid as law of unification and its methodological worldview this chapter points out how it can be universally acceptable. This is true not only for Muslims, but can also fit in with analytical and practical soundness of monetary, financial, banking, and market exchange matters in the global scene. Tawhid as law provides the gateway to all of these and more issues with an endless breadth of intellection and legal command.

Towards Inevitability of Tawhidi Methodological Worldview: Problem with 'Shari'ah-Compliance'

'Shari'ah-compliance' is a human-concocted variant for legitimating product development. Yet it has no legitimate and established groundwork for attaining this goal, except by an institutional juristic imposition (e.g. AAOIFI, OIC, IDB, Bank Negare Malaysia, and Bank Indonesia).

Consider the case of almost all Islamic financial instruments, which altogether are debt instruments. The fatwa of permitting one-third of financial resources to be held in debt is sheer clergy pronouncement that has now permeated into shari'ah ruling. Then there is the entire pricing mechanism of delayed payments (e.g. ijara,. murabaha, takaful, sukuk, and asset valuation). These are all deeply embroiled in subjectivity of rate-setting, there being no rigorous knowledge and model in 'shari'ah-compliance'. The result is a form of riba-setting in asset valuation. The Tawhidi alternative of valuation based on the episteme of unity of knowledge remains unknown.

The Tawhidi methodological worldview formalizes the model of immersing all pricing matters of muamalat, finance, and banking into their complementary interrelationship by inter-causality between these functions and with their inner defining variables. In the absence of understanding the formalism of Tawhidi law of monotheistic unity of knowledge (Qur'anic pairing), all forms of interest rates have remained inextricably embedded in Islamic asset valuation, financial instruments, and thereby in Islamic product development.

'Shari'ah-compliance' is now overloading muamalat matters with fatawa, not recognizing the precision of mathematical and analytical scientific methodology that arises from the Tawhidi law of the Qur'an. The Tawhidi law of epistemic unity of knowledge is transmitted by sunnah, and intelligently discoursed by the learned and devoted Islamic scholars. Contrarily, shari'ah regulations have now embarrassed lock, stock, and barrels the entire body of Islamic economics, finance, and banking. This has caused a state of impossibility with the complications and confusions of 'shar'iah compliance' idea that have befogged the true Islamic worldview.

Generalizing the Methodology and Application of Tawhidi Law Contra 'Shari'ah-Compliance'

The universality and uniqueness of Tawhidi law, though not including in this the hidden attributes (sifat), can resolve the standardization problem of muamalat and Islamic banking caused by 'shari'ah-compliance' fatawa. The positive complementary interrelationship we refer to here is between money, finance, banking, and real economy. We now raise the question: How the debt problem of Islamic financing instruments, Islamic product development, and absence of formalism and applicative methods can be resolved? 'Shariàh-compliance' has not resolved this problem.

Thereby, the inherent pricing mechanisms and product developments have remained mired. For example, 'shari'ah-compliance' continues to commit the dire mistake of focusing on shareholder model of asset valuation as opposed to a stakeholder model along with its extensions beyond the mere five elements of maqasid al-shari'ah. Thereby, 'shariàh-compliance' overly focuses on property rights in its narrow financial sense. The goal of maximization of shareholder's wealth with exogenous ethical excitement as opposed to conscious endogeneity is proferred. Thereby, no analytical substantiation is offered regarding valuation with the conscious nature of endogenous inter-variable relations in the wellbeing function (maslaha), according to Tawhidi law. This is a substantive issue of mathematical formalism and applications.

Wellbeing Criterion

The wellbeing criterion (maslaha) would otherwise comprise the abstraction and applied way of asset valuation, product development, and the associated learned ways of quantitatively evaluating the wellbeing objective criterion. The cardinal, universal, and unique precept of Tawhid as the primal ontological law of monotheistic unity of knowledge could not be embodied in the generality and details of muamalat. This would otherwise comprise the financial world-system of good choices (maqasid), while avoiding the false and questionable ones, even as the evolutionary learning proceeds on discursively (shura).

'Shari'ah-compliance' has fallen deeply into confusion. The non-standardization of fatawa by the various fatawa and fiqh interpretations have gone into quagmire of juristic proliferation. A scientific study of muamalat, finance, banking, and real economy, which Tawhidi methodology offers, is absent in 'shari'ah-compliance' approach.

Consequently, as the global order of intellection is today boldly marching forward with great alacrity into re-examining the epistemological foundations that overarch transdisciplinary areas, so also Islamic economics, finance, banking, real economy and the archaic practice of fiqh and fatawa must give way to the great Tawhidi epistemic questioning. Today's Islamic intellection is confined to innovated ideas of 'shari'ah-compliance'. The abstraction and applications by the Qur'anic monotheistic primal ontology of Tawhid as law and its world-system embedding for the common good must be evaluated by the wellbeing criterion (maslaha). Tawhid as law with its epistemic abstraction and methodological worldview with applications in the Islamic economic, finance, banking, and real economy issues remain forgotten in 'shari'ah-compliance'.

Islamic Product Development by Tawhidi Law of Unity of Knowledge

As another example of contrast between the Tawhidi methodological worldview and the 'shari'ah-compliance' way, we briefly examine the nature of property rights concerning any and everything in Islamic economics, finance, banking, and real economy. The old maqasid al-shari'ah, despite being the blessed gateway towards Tawhidi law of unity of knowledge, has been marginalized due to lack of intellection. What has resulted is a narrow understanding of the five elements of maqasid al-shari'ah as permissible independent states of maslaha (Masud, 1984). Thereby, all such untenable goals as maximization of shareholder's wealth replaced by the maslaha function remain absent. Similarly, all the five elements of maqasid al-shari'ah remain static concepts. No conceptualiztion is done regarding the vista of extensions for wellbeing (maslaha). This would include Muslims, non-Muslims, and the international monetary system in the case of muamalat, finance, and banking. Contrarily, 'shariàh-compliance' has remained distanced by its approach of methodological individualism and archaic postulates of rational choice. Thereby, much of the analytical methodology, as otherwise established by the study of Tawhid as law, has remained distanced, forgotten, and unintelligible to latter days' Muslims. Such Muslims are today working in borrowed mainstream and concocted ideas regarding Islamic everything. Tawhid as law in socio-scientific inquiry presents the unified inquiry of inter-causal relations, and construction of endogenous regulations, policies, inferences, and development of learning institutions. Endogenous intellection in such cases are contrary to Central Bank's exogenously formed regulatory policies and rules.

Riba Problem in Muamalat Explained by Tawhidi Law Contra 'Shari'ah-Compliance'

According to Tawhid as law, product development is secured from riba-like interventions in muamalat, financing, and banking contrarily by the participatory principle of unity between these areas. According to Tawhid as law, even as inter-causality increases between the functions, so also the debt-ridden instruments of 'shari'ah-compliance' are phased out. Conversely, the episteme of unity of knowledge by complementary pairing increases as evolutionary learning proceeds. The result is a true Islamicization (Tawhidization) of muamalat, finance, banking, and real economy along with their endogenous complementary internal variables.

The problem of riba and riba-like pricing of debt-ridden Islamic financing instruments as 'shari'ah-compliance' ruling on this matter is resolved by taking recourse to the Tawhidi organic relational worldview intra-systems and inter-systems with the selected variables. For instance, riba-like pricing is not discounting of futuristic probabilstic expectations according to Tawhidi law of unity of knowledge. Rather, such pricing is done at the 'nearest' point of occurrence of contingencies characterizing various resource flows at these points. At such points the Tawhidi unity of knowledge by systems works in unifying the monetary, financial, banking, and real economy systems. For sustainability, these systems remain interactive, integrative, and evolutionary by learning processes over knowledge and time. Debt problem is thereby avoided when every event point of temporal valuation with contingencies is evaluated by equivalence between money, finance and real value of assets. Islamic banks ought to do this kind of participatory financing and carry out the underlying knowledge dissemination. The central bank ought to take stock of this kind of learning and valuation. The Tawhidi methodological worldview of participative organic pairing (Qur'an 36:36) is now active.

Tawhidi Methodology Contra 'Shari'ah-Compliance' and Kantian Rationalism

Figure 5.1 briefly outlines the contrasting opposites, between Tawhid as the primal ontological law of monotheistic unity of knowledge and the human-concocted notion of 'shari'ah-compliance'. Figure 5.1 explains that reasoning in 'shari'ah-compliance' is discontinuous, as in the case of the Kantian philosophy of rationalism (Friedrich, 1977). Although Kant

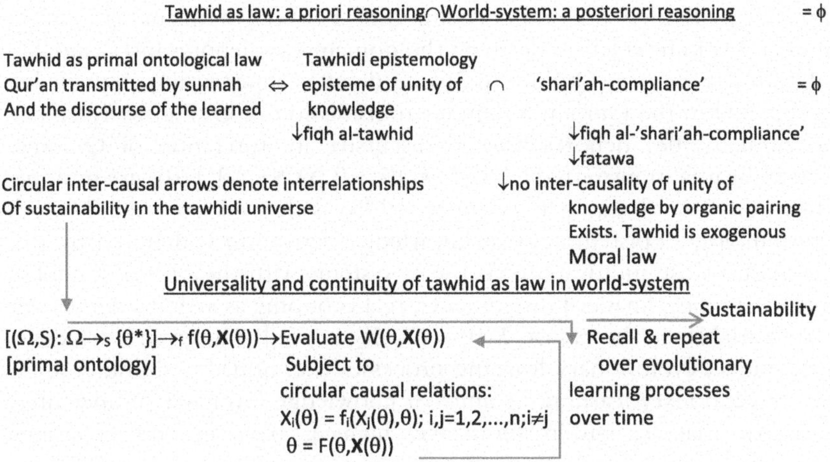

Fig. 5.1 Methodological independence between Tawhid as law and 'shari'ah-compliance'

was a thinker along lines of moral imperative (a priori reasoning), yet his methodology relegated the moral origin of reasoning to the human world-system (a posteriori reasoning). By rationalist reasoning, even though a priori reasoning carries the primacy of moral imperative, yet it remains distanced from the origin of monotheism. Thereby, a priori normative reasoning remains separated from a posteriori positivistic reasoning (this nullity is symbolized by ϕ). Any attempt to unify a priori and a posteriori reasoning in the monotheistic sense failed in Occidentalism. This Kantian way of treating the moral imperative within the dichotomy between God and the unified world-system is repeated in 'shari'ah-compliance'. Thereby the existing approach to the study of Islamic economics, finance, banking, and the needed transdisciplinary inquiry of socio-scientific world-systems with diverse details reflect Kantian rationalism. Morality remained an embellishment of moral philosophy and fails in critical realism (Bhaskar, 1994). So also anything whatsoever that remains independent of and not substantively embedded by Tawhidi law as monotheistic episteme of unity of knowledge is of human innovation. This is how Ibn al-Arabi explained reality entrenched in Tawhid (Chittick, 1989).

Symbols $\cap$ denotes integration commonly between domains. ϕ denotes null and void interaction between the domains. $\Leftrightarrow$ denotes inter-causality of pairing relations (Qur'an 36:36). Ω denotes supercardinal knowledge completed in the Qur'an. S denotes transmission of Qur'an by reference to sunnah. $\{\theta *\}$ denotes learned discursive interpretation of Qur'anic verses (ayaths) transmitted by S; thus, $\{\theta *\} \in (\Omega, S)$. $f(\theta, X(\theta))$ are multiple 'functional ontologies', 'f' (Gruber, 1993) is derived from the Tawhidi epistemology. Epistemology as knowledge derivation is denoted by $\{\theta\}$. $\{\theta\}$ induces the multivariate vector of systems with their inner variables, such as money, financial instruments, real economy as systems defined by their variables in the vector, $X(\theta) = \{X_1, X_2, \ldots, X_n\}[\theta]$ in given interactive, integrative, evolutionary learning processes (i.e. ijtihad = shura, consensus, evolutionary learning in unity of knowledge intra-system and inter-system), that is, within monetary, finance, banking and economic intra-systems and across these inter-systems. W (.) is the conceptual form of wellbeing criterion (maslaha) that establishes the most critical overall unifying effect of Tawhidi law on standardization of rules and methodical applications derived from the primal ontology and its methodology as shown above. The resulting complementarities evaluated and reconstructed (simulated) in terms of the Tawhidi knowledge-induced variables of the wellbeing functional n-dimensional vector are done by the circular causation equations in accordance with the parametrized values of knowledge, $\{\theta\}$. The circular causation equations are denoted by $X_i(\theta) = f_i(X_j(\theta), \theta)$; $i, j = 1, 2, \ldots, n$; $i \neq j$. $\theta = F(\theta, \mathbf{X}(\theta))$ denotes the quantitatively evaluated form of the wellbeing function (maslaha) in the conceptual form, $W(\theta, \mathbf{X}(\theta))$. The primal ontology is continuously recalled intra-systems and inter-systems over continuums of evolutionary learning processes across the historical consciousness of knowledge, knowledge-induced space and time (Qur'an, 76:1). This is the meaning of sustainability of the standardized rules according to Tawhidi law of unity of knowledge (inter-causal pairing, complementarities, participation). The universality, uniformity, and positivistic applications are recursively interrelated with the normative methodology of Tawhid as law.

Summarizing

In conclusion, it is noted that, although there is a claimed derivation of injunctions by 'shari'ah-compliance' towards Tawhid, this does not form a methodological groundwork, except for the use of fiqh and fatawa that support human-concocted implications of 'shari'ah-compliance'. The result thereby, has throughout been its failure to derive the universal and unique foundation as of the Tawhidi ontological, epistemological, and phenomenological worldview of a truly meta-scientific nature in unity of knowledge and the unified world-system. No standardization of fiqh and fatawa is possible in the existing framework of 'shariàh-compliance'.

How will the Tawhidi law be used to gain and sustain standardization of fiqh and fatawa and shari'ah rulings for Islamic banking on appropriateness of instruments, mainly of the secondary financing type, around which product development is carried out? Tawhid as law represents itself in unity of knowledge and the specific issues under discussion in Islamic banks that arise by means of the principle of pervasive organic pairing as complementarities and participations. This feature of Tawhidi rule extends into general-system understanding of the Qur'anic idea of pairing in widening domains of moral-material oneness. Evaluation of the effectiveness of such participatory or complementary processes of adopted rules, instruments, and endogenously derived policies and regulations by discourse is to be done by the estimation and simulation results of the wellbeing function in terms of the inner variables of the objective criterion.

An example is of organic unity of being between money, finance, banking, and real economy, in terms of the functional defining variables and their extensive inter-causal relations signifying the Qur'anic meaning of pairing the universe and its specific details, as in muamalat, finance, and banking. The result is a substantive simulation of the multivariate wellbeing function in the maqasid choices while avoiding the false and unrecommended ones. The resulting fiqh and fatawa in all such universal and unique determination of Tawhidi unity of knowledge by extensive participation as organic pairing signifies inter-system and inter-variable endogenous complementarities in transdisciplinary phenomenology. This approach does not accept differentiation by systemic disjointedness except for unrecommended variables.

Figure 5.1 depicts the human-concocted fiqh and fatawa independently of and not substantively related with Tawhidi law that arises from the learning multiverse (a'lameen) by complementary pairing. The pairing indicator of standardization of issues of Islamic economics, finance, banking, and society is derived from the Qur'an (azwaja kullaha) via sunnah and learned discourse on specific issues. These all reflect the Signs of Allah (ayath Allah).

References

Bhaskar, R. (1994). *Critical Realism*. London, UK: Verso.

Chittick, W. C. (1989). *Sufi Path of Knowledge*. Albany, NY: State University of New York Press.

Choudhury, M. A. (2006). *Koran and Science* (with Generic Titles). 5 vols. Lewiston, NY: Edwin Mellen Press.

Choudhury, M. A. (2014). *Tawhidi Epistemology and Its Applications (Economics, Finance, Science, and Society)*. Cambridge, UK: Cambridge Scholars Publishing.

Choudhury, M. A. (2016a). *Heterodox Islamic Economics: The Emergence of an Ethico-Economic Theory*. London, UK: Routledge.

Choudhury, M. A. (2016b). *The Absolute Reality in the Qur'an*. London, UK: New Palgrave Macmillan.

Choudhury, M. A. (2017). *A Phenomenological Theory of Islamic Economics*. Kuala Lumpur, Malaysia: University of Malaya Press.

Friedrich, C. J. (Trans.). (1977). *The Philosophy of Kant: Immanuel Kant's Moral and Political Writings*. New York, NY: The Modern Library.

Gruber, T. R. (1993). A Translation Approach to Portable Ontologies. *Knowledge Acquisition*, *5*(2), 199–220.

Masud, M. K. (1984). *Islamic Legal Philosophy: A Study of Abu Ishaq al-Shatibi's Life and Work*. Islamabad, Pakistan: Islamic Research Institute.

CHAPTER 6

Tawhid as Law Contra 'Shari'ah-Compliance' in Malaysian Advisory Commission

Introduction

The consequences of the contradictions between the so-called shari'ah-compliance implications and its ambivalence of Tawhid in worldly actions lead to unauthentic inferences in the shari'ah. The inconsistency is also found in the case of oppositeness between the functional meaning of maqasid al-shari'ah and shari'ah-compliance. Indeed, even maqasid al-shari'ah is not *the* Islamic law. It is incomplete, and shows vagaries caused by human design, urf, and adah (customs and cultures). It also presents a traditional perspective of worldly affairs (muamalat). It ignores to study the grand design of universal unity and evolutionary learning dynamics as systemic relational worldview (Auda, 2008).

On the other hand, there are the short-run quick rulings according to the so-called shari'ah-compliance approach. Shari'ah-compliance by its short-run nature has avoided the permanently great questions of wellbeing, its extendibility and sustainability, and circularly integrated deductive-inductive unity of reasoning. Thereby, the inferences drawn from shari'ah-compliance are cursory and referentially traditional in nature. Thus, neither maqasid al-shari'ah nor shari'ah-compliance conveys any

---

A version of this chapter appeared in *International Journal of Law and Management, 63:6*.

M. A. Choudhury, *Tawhid and Shari'ah*,
https://doi.org/10.1007/978-3-030-49087-4_6

aspect of the true Islamic law.[1] Besides, they do not give the pivotal, most indelible, unchangeable nature of the primal ontological law of Tawhid. Yet it is true that the shari'ah can reflect the understanding of Tawhid in worldly matters to the extent of human involvement in this quest. Yet the shari'ah does not represent the res extensa and res cogitans vista of the Qur'an mapped as this must be by the authentic sunnah as the anthropic explicatory medium of the unity of the world-system in its generality and details (Choudhury, 2015a).

Maqasid al-shari'ah once derived and permanently linked to the investigation revolving around Tawhid in all functional matters becomes only a secondary sustained and continuous system of juristic deliberations. In this state of the shari'ah in its incompleteness and anthropomorphic nature of long-term continued deduction and induction, reasoning leads to discourse. Discourse is characterized by ijtihad (investigation), qiyas (analogy), and ijma (consensus). But even so among different and often contending Muslim sects, there has resulted non-harmonization and non-standardization of issues regarding the acceptability of certain financing instruments. On the other hand, rules derived for financing instruments are very often based on untenable conclusions arising from traditions while avoiding the intellection into new vistas of knowledge arising from the Qur'an and the authentic ahadith (sayings of the Prophet Muhammad, and thus sunnah) (Asad, 1987). That is, Tawhid as the primal ontological law of unity of knowledge between diversities of being and becoming in the grand design of the universe, in its conception of generality and particulars, has been forgotten for long now (Buchman, 1998; Bruteau, 1997).[2] Therefore, for instance, the shari'ah has never been in practice in history except by the mandates of rulers. Contrarily, Tawhid lives forever.

[1] Qur'an (10:66): "Unquestionably, to Allah belongs whoever is in the heavens and whoever is on the earth. And those who invoke other than Allah do not [actually] follow [His] 'partners.' They follow not except assumption, and they are not but falsifying".

[2] Buchman, D. writes (p. xiii): "It is not enough to abstract God from the world in the manner of the theologians; one must also perceive His presence. The Qur'an often refers to its teachings about God by employing the words mathal and mithal, 'similitude' and 'likeness', so it is not surprising to see that al-Ghazali pays a good deal of attention to explaining the true Qur'anic sense of these words. Certainly he is striving, among other things, to show that the 'symbolism' of the Qur'an should not be thought of primarily as literary imagery, as similes and metaphors. On the contrary, God employs the language that He employs to clarify the actual nature of reality. At issue are the structure of the cosmos and the human soul."

Bruteau (p. 179) writes: "If you can see the God you love present in, even as, this world, then feel that union and rejoice in that. And be active in it, contribute to it, participate in the

Objective

Scholars writing in the field of a critical examination of the shari'ah have pointed out the partial equilibrium and partial-system orientation of the existing Muslim understanding of the shari'ah as a partial worldview (Weeramantry, 2001; Hassan, 2002; Coulson, 1984). Indeed, the shari'ah fell into such dysfunctional problems by the intrusion of the traditional interpretive (fiqh) approach made by the traditionalists in denial of the holistic intellection process of Tawhid. The result has been a heavily personified development of administrative impositions, often based on opinion, rather than the natural law of justice and balance (mizan) as in the Qur'an, and being mapped onto the generality and particulars of the world-system by the sunnah and authentic discourse.

The Overshadowing Problems of Social Compliance Idea and the Tawhidi Rejection Concerning Islamic Finance

In this chapter, we will consider examples of the (i) confusing and untenable intertwining between financing instruments in shari'ah-compliance rulings. (ii) We will then follow up this issue with the financial implications of the Tawhidi methodological worldview of unity of knowledge. We start by fixing the premise of Tawhid as the law of 'everything' on the goal of wellbeing. Wellbeing as the objective criterion of unity of knowledge is evaluated in reference to the inter-causal relations between representative choice variables. The consequential implications are regarding (iii) relational organism in Tawhidi methodology versus 'uqud or contractual nature of static financing instruments in shari'ah-compliance; (iv) general and particular derivations of Tawhidi methodology in complexity by richness versus linear aggregation of functions in both maqasid al-shari'ah and shari'ah-compliance; (v) the comprehensive intellection in Tawhidi methodological worldview contrary to institutional and juristic perspectives of shari'ah-compliance contracts. (vi) The way to attain such goals is the formalism and application of the objective criterion of wellbeing in Tawhid, which is analysed subject to a system of inter-causal relations between critical variables characterizing the problem under study. In a special case, such

building, in the artwork, in the healing, in the understanding. This is where Reality is. You yourself are both a member of the Finite and a member of the Infinite."

a perspective of wellbeing criterion with systemic relations of inter-causality between the choice variables in terms of the Tawhidi epistemology of unity of knowledge can be formalized by a vector of financing instruments. (vii) Such a vector of variables can be extended to the inter-systemic case of multistage sectoral-variable systems that are complemented with the finance sector. Examples of such complementary sectors with the financial sector are the economic, social, and socio-scientific sectors of investigation.

The Tawhidi Methodological Outlook Contrary to Shari'ah-Compliance Approach

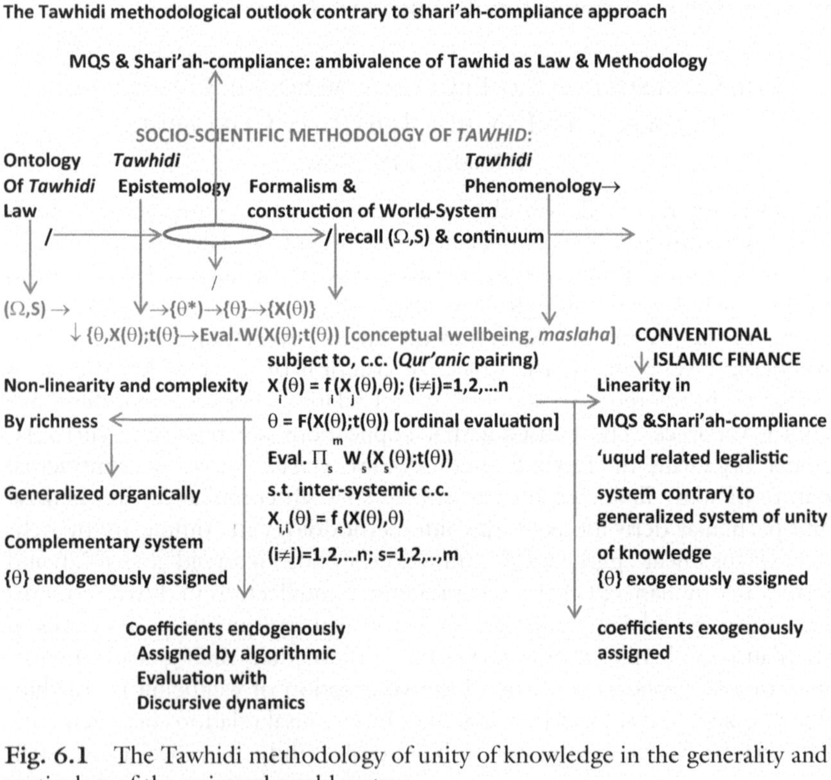

Fig. 6.1 The Tawhidi methodology of unity of knowledge in the generality and particulars of the universal world-system

Deductions Concerning Selective Shari'ah Financing Instruments

A principal problem with the 'aqd-(contractual)-based financial deductions are regarding rent valuation over time. This constitutes the complex yet rich reality of complex compounding of factors caused by events of intertemporal nature, such as the point 'h' on the historical path of asset valuation shown by HH in Fig. 6.2.[3] The Tawhidi worldview considers the point 'h' to be embedded in the knowledge, space, and time dimensions by virtue of its probabilistic nature over time. We therefore denote such a point on life trajectory of everything under valuation by $h(\theta)$ and $HH(\theta) = \int h(\theta)d\theta$, and its multiple integrable complex of variables and their relations.

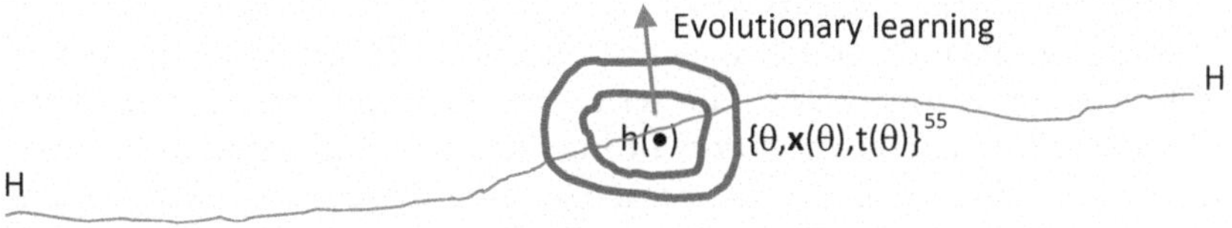

Fig. 6.2 Tawhidi events in knowledge, space, and time dimensions

[3] An analytical function defined by its differentiability and continuity properties can be expanded in the small neighbourhood of $h(\bullet)$ by Taylor's series as follows (Jeans, 1970): Let $f(\mathbf{x}(\theta))$ denote a continuously differentiable function (any kind of financial benefits denoting asset value) in the knowledge, space, and time dimensions for a specific point of time $t(\theta)$, and thus at $(\theta, \mathbf{x}(\theta), t(\theta))$. Taylor's series in the neighbourhood $h(\bullet)$ around the expected value (mean), $E(\mathbf{x}(\theta))$, is written as,

$E(f(\mathbf{x}(\theta))) = f(E(\mathbf{x}(\theta))) + [(\mathbf{x}(\theta) - E(\mathbf{x}(\theta)))] \times f'(\mathbf{x}(\theta)) + [(\mathbf{x}(\theta) - E(\mathbf{x}(\theta)))]^2 \times f''(\mathbf{x}(\theta)) + \ldots + [(\mathbf{x}(\theta) - E(\mathbf{x}(\theta)))]^n \times f^n(\mathbf{x}(\theta)) + \ldots \infty$ termsfor, $\text{plim}\{\theta\} = \theta \in (\Omega, S)$. (Ω, S) denotes the primal ontology of Tawhid and the occurrence of events being at each point of time. Thus, Taylor's expansion exists everywhere within $h(\bullet)$ and across $HH(\theta)$.

Because $\mathbf{x}(\theta)$ denotes a vector/matrix and so on of inter-causal variables, any positive monotonic transformation of this vector in functional form is equally acceptable. We chose this functional form as the wellbeing function, $W(\theta, \mathbf{x}(\theta))$, as a functional to evaluate the 'as is' and the 'as it ought to be' states of inter-variable complementarities. Such complementary 'pairing' is the evidential sign of Tawhidi unity of knowledge as organism. With the resulting statistical simplifications into a quadratic form, we can write the above expression as,

$W(\theta, \mathbf{x}(\theta), T) = \int_0^T \int_0^{x(\theta)} \int_0^{\theta} a_t(\theta)^* \left[W\left(E\left(\mathbf{x}(\theta)\right)\right) + b_t(\theta)^* \text{Var}\left(\mathbf{x}(\theta)\right)\right] \mathrm{d}\theta \mathrm{d}\mathbf{x}(\theta) \mathrm{d}t$, encompassing the knowledge, space, and time dimensions.

$h(\bullet)$ - embedding in knowledge, the $\{\theta\}$ - domain, means it is interrelational with many similar points in this dynamic domain. This is the systemic and cybernetic nature of the $h(\bullet)$—neighbourhood, denoted by the expansionary domain as shown.

Next, in reference to Fig. 6.1, every such $h(\bullet)$ embodies the space variable denoted by $\{\mathbf{x}(h(\bullet))\}$.

We thereby have a tuple embedding to denote the point, $h(\bullet)$. This point is denoted by $\{h(\theta), \mathbf{x}(h(\theta))\}$. In the general case for all points in the $h(\bullet)$—neighbourhood we write, $\{\theta, \mathbf{x}(\theta)\}$. The bold symbol $\mathbf{x}(\theta)$ can denote vector, sequences of vectors, matrices, and extended mathematical forms and relations (Gel'fand, 1961).

The time variable enters as a record-keeping element of events with the probabilistic occurrence of $\{\theta, \mathbf{x}(\theta)\}$. We thereby write the $h(\bullet)$-point in regards to intertemporal continuity sustained by $\{\theta\}$ as the non-Cartesian learning coordinate, $\{\theta, \mathbf{x}(\theta), t(\theta)\}$.

Financial valuation occurs at every point along $HH(\theta)$ at specific points of time, and across $HH(\theta)$ in terms of intertemporality. The point $h(\{\theta, \mathbf{x}(\theta), t(\theta)\})$ carries the forces of interaction and integration by algorithmic and discursive experience across multidimensional systems and cybernetics defined over knowledge, space, and time.

Tawhidi Financial Valuation

Consider three h's$(\bullet)$: house insurance, car insurance, and life insurance (all under takaful, insurance). These securities interact to set up a unified predictable but probabilistic scenario. For instance, death and life expectancy can affect life insurance payments, and thereby payment of house insurance and car insurance. Likewise, payment of car insurance affects payment of house insurance. That is why combined house insurance and car insurance results in lower premiums. Such a combined insurance payment affects the volume of life insurance payment over the lifetime. Thus, at the end, the Tawhidi perspective is to set up a social wellbeing function to test and simulate for positive complementarities between the three forms of insurance premiums, subject to circular causation interrelations between them. The general model is shown in Fig. 6.1.

Contrarily, in the case of shari'ah ruling (Shari'ah Advisory Council (SAC), 2016), each of the above-mentioned premium relations can be taken within the hire-purchase sale contract. In shari'ah-compliance idea, each premium rule is treated as a separate 'aqd (contract). For instance, the ruling is that two risky contracts cannot be combined for the sake of financial valuation. Such 'aqd-ruling differentiates life insurance to be independent of its cause and effect in relation to the other premiums. For instance, the car insurance is shown as not being related with the other premiums via the probability of accident. House insurance is seen independently of the ability to pay premiums on the car insurance and the life insurance, on grounds of ability to pay.

Following is the shari'ah enunciation (SAC, op cit) regarding car insurance (AITAB).

Car Insurance, AITAB: Shari'ah-Compliance Rulings and Contradictions

1. The modus operandi of AITAB shall consist of two independent contracts, namely ijarah financing contract (rent) and al-bai' (sales) contract. Thus, the joint and interactive nature of lease to purchase is not an interdependently structured contract for car purchase, and thereby car insurance with other ones.
2. The AITAB agreement shall include a clause that specifies, "will purchase the vehicle" at the end of the lease period, as well as a clause on early redemption by the lessee.

   This contract ought to be structured as a conditional probabilistic one based on contingencies at the beginning of the contract and the end of contract.
3. The deposit paid to the vehicle dealer does not form a sale contract since it is deemed as a deposit that has to be paid by the Islamic financial institution.

   Yet the Islamic financial company will include this deposit in his total price and will be subject to probabilistic contingencies.
4. In line with the principles of ijarah (rent), the Islamic financial institution as the owner of the asset shall bear all reasonable risks relating to the ownership.

   How can this indemnify the purchaser of the asset from the probabilistic contingencies? The owner is free!

Home Insurance, Commodity Murabaha (Upfront Mark-Up Financing by Financial Company)

In recent times, murabaha (mark-up pricing) as a primary financing instrument and several secondary financing instruments are all debt instruments. The debt is charged by exogenously set rental rates (e.g. ijara rates) between the contracting partners on outstanding debt. This mode of financing has been criticized as a kind of riba (interest)-financing over time (Misri, 1987). Riba is strictly forbidden in Islam on all matters. The Islamic implication is serious in the intertemporal sense; and there is no meaning of debt momentarily. Intertemporal charging of such prefixed ijara rates over time contaminates the legitimacy of ownership and distribution of assets and their benefits with shadow rates of riba. In the presence of all forms of riba, the principle of wellbeing (social benefit), called maslaha, is destroyed at every moment and event like $h(\bullet)$ along $HH(\theta)$ in Fig. 6.2 across knowledge, space, and time dimensions.

The pursuit of shari'ah-compliance has failed to think about such real intricacies of rate-setting that remains central to ijara-type periodic payments on all shari'ah instruments concerning debt. See appendix on the topic of Islamic financing irrelevance to Modigliani and Miller's optimal structure of the financial firm (Jeans, 1970). This Islamic financial irrelevance further proves that presently no model of rate determination has been found in the shari'ah, which would otherwise evaluate the Islamic valuation of financial events at $h(\bullet)$ over time.

Bank Notes Based on Ijara Concept in Shari'ah-Compliance

In this regard, Malaysian Shari'ah Advisory Council unravels the inherent problem, as in its concept of bank notes based on the ijara concept. This financing method works in the following way: (i) Bank Negara Malaysia (National Bank of Malaysia) will sell the beneficial interests of its real property (such as land and building) to a Special Purpose Vehicle (SPV). The SPV will then lease the property to Bank Negara Malaysia for a specific period of time through execution of an ijarah muntahia bi al-tamlik agreement. This is a lease contract ending with acquisition of sale or gift. As a consideration, Bank Negara Malaysia will pay rent (at a rate which is agreed upon during the conclusion of the contract) for every six months throughout the lease period; (ii) Bank Negara Malaysia will provide to the SPV wa'd (rent to own) to repurchase the property at an agreed price on the maturity date. The idea is too cluttered. It can be immensely simplified.

In these contractual agreements, the ijara rate and the future price are set exogenously by partnership contracts and not by socio-economic and wellbeing realities (maslaha realities) arising from market transactions over time and by changes in relations and through discourse, as is the case with the Tawhidi knowledge flow denoted by $\{\theta\}$.

Commodity Murabaha Contract

In the case of debt liquidation payment on a commodity murabaha contract, a similar problem of rate-setting continues and deepens because of the mega-projects that get involved. The ijara rate now equals the rent charged on an asset, which in turn includes a liquidate rate + fair profit margin + exigency rate. The spread of this total rate-setting takes place at every point like $h(\bullet)$ over time. Yet in such rate-setting none of the premium or rent-setting is an actual rate around the neighbourhood of $h(\bullet)$ in the 'nearest' probabilistic sense of valuation, which is the objectified probability measure of subjective probability based on socio-scientific algorithmic computation and expert discourse of the shura (consultation and discourse) type (O'Donnell, 1989; Lawson & Pesaran, 1985; Choudhury, 2015b).

Generalized Inference Regarding Rents on Shari'ah-Compliance Debt Instruments

Because of the problem of rate-setting of the ijara type, which in fact runs throughout all the Islamic financing instruments as debt instruments, there does not exist a viable and meaningful interest-free approach to asset valuation in the present ruling under shari'ah-compliance. Firstly, there is no model in existing shari'ah-compliance literature that can do safe valuation of assets by rate-setting, like the intergenerational valuation model of assets and benefits by the 'nearest' type objectified probability measures (Choudhury, 2011).[4] Secondly, the exogenous rate-setting appears as shadow rate of interest in shari'ah-compliance idea. Thereby, the intercausal relations between housing premiums (rents) in murabaha and car

[4] Use the Newton-Rhapson method (Kelley, 2003) of calculating iterative roots of 'r' of the terminal-value polynomial within an interval of the 'nearest' point of a simulated value denoted by ε in (0,1). Let the overlapping generation asset valuation polynomial function be given by the terminal-value formula, $TV(.) = -I + \Sigma A_t(1 + r(\theta))^t$. Let I denote investment outlay. $A_t(\theta)$ denotes cumulative asset value over time 't' as valued by the social wellbeing evaluation at each selected point of time. At a simulated 'nearest' rate of return, $\Delta r(\theta) = dr/d\theta = \varepsilon(\theta) \in (0.1)$, after mathematical simplification, the Newton-Rhapson iteration method yields [θ is repressed],

$$\varepsilon = TV/(dTV/d\theta) = [-I + \Sigma A_t(1 + r(\theta))^t]/[\Sigma t.\ A_t(1 + r(\theta))^{t-1}.\ (dr/d\theta)] = [-I + \Sigma A_t(1 + r(\theta))^t]/[\Sigma\{t.\ A_t(1 + r(\theta))^{t-1}\}.\ \varepsilon].$$

Thereby, we can write this equation in the computable form, δ's being functions of (A's, ε, θ).

insurance premium in AITAB do not exist. The deferred rate-setting ad hoc contaminates both of these transactions by the formula of 'aqd contractual legalization of exogenous price-setting mechanism.

The riba (interest) nature of rate-setting under shari'ah-compliance approach of 'aqd legislation causes at least two further serious illegitimacy problem in Islamic financing. Firstly, the debt financing nature of all Islamic financing instruments is caused by the lack of understanding of the method of valuation in the money, finance, and real economy (MFRE) circular causation relations. This vast topic can be only summarized here by stating that the value and quantity of monetary transmission must be equivalent to value of spending in the good things of life in the real economy. The result then is the establishing of the circular causation relations between money, finance, and real economy. This phenomenon is of an intertemporal nature across the historical trajectory $HH(\theta)$, realizing itself at every point $h(\bullet)$, as in Fig. 6.2.

Consequently, as in the case of murabaha and its like in ijara financing, such as istisna (manufacturing loan to complete a contracted job), bithmanagil (deferred payment sale), tawarruq (cash murabaha), sukuk (Islamic bond) and so on, the delay in payments in the MFRE model exists permanently. Contrarily, in the Tawhidi interrelational model of unity of knowledge, the monetary and resource mobilizations are recycled and regenerated through the imminent 100 per cent reserve requirement monetary system that drives the circular inter-relations. All that is required in the Tawhidi perspective of MFRE model is the well-functioning of a life-fulfilment regime of development to prevail in tandem with the 100 per cent reserve requirement monetary system (100%RRMS). The 100%RRMS may exist with the gold standard as a numeraire. But this standard need not necessarily be a circulatory gold currency.

The end result in the MFRE arrangement then is this: Islamic bank savings are continuously mobilized into the life-fulfilment regime of real economy (i.e. not withheld to create debt) by participatory financing instruments. Examples of such instruments are co-financing, joint venture, equity financing, profit–loss sharing, productivity-based exchange rates and yields on regenerated asset values. The central bank adjusts its marginal supply of money to make the equation on quantity and

$\delta_0(\varepsilon) + \Sigma_t\delta_t(\varepsilon)(1 + r)t = 0$. This equation can be solved for 'r' for assigned simulated choices of ε-values in the interval (0, 1), given discoursed wellbeing 'θ' values at the 'nearest' points of asset valuation along $HH(\theta)$, as in Fig 6.2.

valuation in MFRE model uphold at all levels of resource mobilization into recommended spending.[5] Thereby, bank savings are annulled. Hence, debt created by an inequality in such an otherwise mismatch in the MFRE model is permanently cancelled.

Debt permissibility in shari'ah-compliance is thus a misconception. Debt is logically inadmissible in response to the maslaha principle generated by the holding of Shatibi-basket in the MFRE model, according to the Tawhidi methodology of inter-causal organic relations of unity of knowledge with zero interest rate of any kind.

The other serious problem of shadow rate of interest caused by the exogenous 'aqd treatment of ijara rates intertemporally is its riba contamination across all the so-called shari'ah-compliance instruments based on debt. This is true between AITAB and murabaha in terms of mark-up financing as debt payment by ijara rate. Thereby, their inter-causal relations are also adversely affected by this illegitimacy of the shadow rate of interest as ijara rate. The implication is that this spread of adverse effects distorts the wellbeing criterion maslaha, by the interruption in the positive complementarities between the essentials of just value and balance (al-wasatiyyah). This goes against the Tawhidi ontology of pervasive inter-causal unity of knowledge between the good things of life. The shari'ah-compliance problem is caused by the lack of contextual understanding and the knowledge of appropriate method of rate-setting under 'aqd-based fatawa (ruling).

Life Insurance Premium

Life insurance is not disapproved under either the shari'ah or the Tawhidi law.[6] Rather, it is the method of financing and premium setting that poses the takaful problem in general. One notes that to date there is no truly methodological approach in Islamic financing for rate-setting in the absence of riba. Rate-setting involves setting of premiums; mobilizing such premiums through appropriate Islamic channels; affordability of

[5] Qur'an (2:268): O you who have believed, spend from the good things which you have earned and from that which We have produced for you from the earth. And do not aim toward the defective therefrom, spending [from that] while you would not take it [yourself] except with closed eyes. And know that Allah is Free of need and Praiseworthy.

[6] Qur'an (49:10): "The believers are but brothers, so make reconciliation between your brothers and fear Allah that you may receive mercy."

caring; and thereby attaining social wellbeing. These are long-term sustainability attributes in general. But they have short-term probability of incidence of mortality rates and contingencies. A substantive difference between mainstream premium setting and takaful insurance is the role that interest rate plays in pricing contingencies. Yet the long-term structure of valuation of premiums remains critical in both of these kinds of insurance contingencies.

For instance, consider the general usage of valuation formula in the neighbourhood of $h(\bullet)$ given by Taylor expansion of continuously differentiable function as the premium valuation function:

$$E(f(\mathbf{x}(\theta))) = f(E(\mathbf{x}(\theta))) + [(\mathbf{x}(\theta) - E(\mathbf{x}(\theta)))] \times f'(\mathbf{x}(\theta)) + [(\mathbf{x}(\theta) - E(\mathbf{x}(\theta)))]^2 \times f''(\mathbf{x}(\theta)) + \ldots + [(\mathbf{x}(\theta) - E(\mathbf{x}(\theta)))]^n \times f^n(\mathbf{x}(\theta)) + \ldots \infty \text{terms}. \quad (6.1)$$

This expression is rewritten from the insured's point of view by considering,

$$E(\mathbf{x}(\theta)) = \sum_{j=1}^{m}\sum_{i=1}^{n} \text{Prob}_j^* \mathbf{x}_{ij}(\theta). \quad (6.2)$$

Expression (6.2) gives the expected value of '*i*' number of securities (insurances purchased) under *j*-contingencies. In the simple case of a flat amount of insurance purchased for *n*-cases with different probabilities of occurrence of *j*-contingencies, we write,

$$E(\mathbf{x}(\theta)) = n^* \sum_{j=1}^{m} \text{Prob}_j^* \mathbf{x}_j(\theta) / n = \sum_{j=1}^{m} \text{Prob}_j^* \mathbf{x}_j(\theta). \quad (6.3)$$

$$\text{In the case of life insurance, Prob}_j = (1 - I_{y+t} / I_y), \quad (6.4)$$

with I_y denoting living cohort of insured at the age of y; I_{y+t} denotes the same at the age of $(y + t)$; (I_{y+t}/I_y) is the probability of survival; $(1 - I_{y+t}/I_y)$ denotes the probability of death.

Now, in the conventional case of insurances purchased by $\{\mathbf{x}_{ij}(\theta)\}$ as corresponding premium payments, the expression $\mathbf{E}(\mathbf{x}(\theta))$ would be the present discounted value of future expected premium payments (purchasing insurances). The discount factor is the time preference rate, which is simply the time value of money. This rate is equivalent to the nominal effective interest rate(s), according to the time of occurrence of events in the future and the opportunity cost measuring perception of occurrence of future contingency. Robinson (1961) pointed out such a time-based notion of opportunity cost as an incorrect one.

In the case of takaful, no premium calculation has been made to date. Hence, it is replaced ad hoc by certain shari’ah rulings relating to ways of premium calculation on security and arrangement of insurance claim. Consequently, shari’ah boards have disallowed the two kinds of claims and payments like security and premiums. These modes of security financing are replaced by Rahn (mortgage) and musahamah, meaning periodic payments to and claims from takaful companies by the insured. The debt problem of ijara rate-setting persists.

The Tawhidi principle contrarily is that if financial flows under actual contingency (‘nearest’ point overlapping generation valuation) serve the wellbeing of insurance pooling partners and society at large, as by way of just payment (jus pretium) for a legitimate and actual cause of claim, then such a fair transaction could be permissible.

Now from the viewpoint of wellbeing based on the complementary and fair business dealing between paying partners and obligations of takaful as insurance companies, the above equation is written as follows:

$$E\left(W\left(\mathbf{x}(\theta)\right)\right)=\sum_{j=1}^{m}\sum_{i=1}^{n}\text{Prob}_j^{*}W_{ij}\left(\mathbf{x}_{ij}(\theta)\right). \tag{6.5}$$

As shown above, this equation simplifies to the following one:

$$\mathbf{E}\left(W\left(\mathbf{x}(\theta)\right)\right)=\sum_{j=1}^{m}\text{Prob}_j^{*}W_{ij}\left(\mathbf{x}_j(\theta)\right)=\sum_{j=1}^{m}\text{Prob}_j^{*}W_{ij}\left(\mathbf{x}_j(\theta)\right). \tag{6.6}$$

In the case of business insurance, and therefore life, $\text{Prob}_j = I_j/I_0$, with I_0 denoting total pooled outlay in insurance by the insured participants as paid-up capital. I_j denotes the claim against contingency j; $I_j < I_0 \cdot (I_j/I_0) < 1$ is the probability of a claim in specific types of contingency.

The difference between the takaful wellbeing objective and either the conventional premium capitalization or the silence of shari’ah on capitalization methods is that there cannot be a time preference form of capitalization rate in Islamic valuation formula. Instead, valuation is done 'nearest' to the points of occurrence like $h(\bullet)$ as was explained earlier. 'Nearest' to every such point, complex systems of interactive, integrative, and evolutionary inter-variable causal relations arise in respect of the embodiment of Tawhidi ontology of unity of knowledge in the objective criterion of wellbeing valuation at such points.

The Wellbeing Objective Criterion

The Wellbeing Function in the Tawhidi methodological context is the objective criterion to study the degree to which inter-variable complementarities exist in the light of choices resting upon the precept of organic unity of knowledge between the blissful possibilities pointed out by maqasid al-shari'ah. Wellbeing function is firstly a theoretical abstraction defined in terms of Tawhidi unity of knowledge. This in turn is explained by circular causation between organically interdependent maqasid al-shari'ah choices defined by their specific variables. The degree to which such positive complementarities exist between the variables or can be constructed out of simulacra of the coefficients of the relational model results in inferences on sustainability and social balance explained by such interrelations and simulated choices of coefficients. Thus, at the end, through the medium of wellbeing, the abstraction and the quantitative estimation of wellbeing bring together the three precepts. These are, namely, the primal ontology of Tawhid as the law of unity of knowledge, the gateway of maqasid al-shari'ah to blissful choices; and the actualization of balance and sustainability by the principle of al-wasatiyyah (Choudhury, 2016).

The probabilistic wellbeing equation given above is fully estimated by the underlying circular causation relations of the $\{\mathbf{x}(\theta)\}$ - vector at all points of the $HH(\theta)$ trajectory of asset valuation and rate-setting over knowledge, space, and time dimensions. Besides, each point like $h(\bullet)$ has the microscopic structure. These issues of substantive evaluation are not found in the concept of shari'ah-compliance. Hence, shari'ah-compliance method does not yield an authentic methodology of selection of valuation rules of sales, contract, and financing, or anything like the Tawhidi socio-scientific worldview.

The Portfolio Financing Model in the Light of Tawhidi Methodology

According to Tawhidi methodology, the above-mentioned three areas of insurances for rate-setting are interactively combined together in terms of their various composite variables. The area of shari'ah-compliance does not treat this aspect of the problem of rate-setting as is the case of Tawhidi methodology of organic unity of knowledge. Independent 'aqd financing contracts of shari'ah-compliance cannot construct an interdependent portfolio model of interactive financial instruments. Therefore, Islamic financing under shari'ah-compliance approach can at best rest on mainstream type modelling. This is equally the permanent feature of both mainstream financial portfolio theory and of present days' formulations in Islamic finance.[7]

Islamic Input-Output Portfolio Theory Under Tawhidi Circular Causality of Unity of Knowledge

Here therefore is a formal model of interdependent and inter-causal financial portfolio. We set up the interaction table between securities of the portfolio as follows: The imminent portfolio model with θ-embedding

[7] An example is the capital asset pricing model stated as follows: An investor's total outlaid fund is divided in a diversified stock market between risky market portfolio of securities and risk-free securities. The market portfolio is always risky notwithstanding whatever is the degree of diversification of the systemic risky market portfolio. We denote the total rate of return (r) from investor's investment in such a divided portfolio market in terms of market rate of return, r_m (systemic risk, e.g. equity rate) and r_f (risk-free portfolio, e.g. interest rate). The CAPM formula of the diversified asset rate of return is then expressed by the equation, $r = r_f + \beta(r_m - r_f)$.

Today, every Islamic finance academic misconstrues this formula for Islamic financial engineering. For consider the cultural ingredient of the non-Islamic investor in the absence of social consciousness as would arise otherwise from the Islamic transformation premised on the Qur'an and the sunnah (Ω, S). This is that, interest is forbidden. Yet interest is permanent in CAPM by its risk-free component. On the other hand, investor's dissonance between interest-free interest and risky market yields will not dissuade the investor from buying government bonds, whose yield is a shadow rate of interest and a way of financing unnecessary government sponsored public projects.

The incorrectness of CAPM was pointed out in the literature by the presence of outliers removed away from the CAPM straight line fit (β-fit). This shows the unreliability of 'β' for predictability using the CAPM. CAPM characterizes the culture of non-Islamic investment behaviour. It can be replaced only by a new brand of Islamic investment behavioural model. This is explained here by the consequential Islamic financial outlook on risk and return. The inter-variable embedding by induction of θ-values now becomes indispensable.

and inter-variable circular causation relations is an adapted form of the input-output model, particularly adopted to interdependent portfolio security-specific inter-causal relations. We show below that the input-output tabular expression results in a form of evaluation of the wellbeing function, subject to the system of circular causation relations between the critical variables of the input-output model of the inter-causality-related portfolio.

Input-Output Adaptation of the Islamic Financing Portfolio Model

Let X_i^k denote the total investible fund in ith Islamic security (instrument) in the presence of *k*-contingencies.

Let x_{ij}^k denote the inter-security (inter-instrument) flow of funds between (*i*, *j*) securities (instruments) from the total investible fund with *k*-contingencies. Such inter-variable flows uphold their explanation in terms of the inherent θ-embedding. Such inter-flows of funds are contrary to ‘aqd-based financing of shari’ah-compliance. The presence of contingencies makes all fund flows to be probabilistic in nature. This state in turn conveys the continuously inherent property of interaction, integration, and evolutionary learning, which is an essential character of the Tawhidi wellbeing function as defined (maslaha).

Let x_i^k denotes the residual, if any, from the outlay of funds X_i^k after all inter-security (inter-instrument) distributions have been made under the occurrence of k-contingencies.

Prob$(x_{ij}, x_i)^k$ denotes joint probabilities of occurrence of (x_{ij}, x_i) at *k*th contingency. It occupies the following conditional form: Prob$(x_{ij}, x_i)^k$ = Pr ob$(x_j | x_{ij})^k$. Prob$\left(x_{ij}^k\right)$ and its further breakdown into more composite conditional probabilities between the variables as shown (Hogg & Craig, 1965).

Each of the variables and probabilities is induced by θ value (not shown), which carries the entire methodology of unity of knowledge derived from the Tawhidi epistemic roots, and into the construction and evaluation of the expected wellbeing function, subject to its circular causation relations between the variables. Enumeration of variables is by, *i*, *j* = 1, 2, …, *n*; contingencies are enumerated by *k* = 1, 2, …, *m*.

$$\begin{array}{cccccc} & X_1^{\ k} & X_2^{\ k} \supset & X_n^{\ k} & \text{residual } x \\ X_1^{\ k} & x_{11}^{\ k} & x_{12}^{\ k} \supset & x_{1n}^{\ k} & x_1^{\ k} \\ X_2^{\ k} & x_{21}^{\ k} & x_{22}^{\ k} \supset & x_{2n}^{\ k} & x_2^{\ k} \\ . & . & . & . & . \\ . & . & . & . & . \\ X_n^{\ k} & x_{n1}^{\ k} & x_{n2}^{\ k} \supset & x_{nn}^{\ k} & x_n^{\ k} \end{array} \tag{6.7}$$

We can derive the system of equations arising from the probabilistic form of entries in the input-output table.

$$X_i^k = \sum_j x_{ij}^k \cdot \text{Prob}\left(x_{ij}^k\right) + x_i^k \cdot \text{Prob}\left(x_i^k\right). \tag{6.8}$$

The probabilities are conditional ones; and the $\left(X_i^k, x_{ij}^k, x_i^k\right)$ appear in their expected forms.

By the method of input-output modelling, we transform the above table into its probabilistic coefficient form induced by θ value (not shown). That is,

$$a_{ij}^k = \frac{x_{ij}^k}{X_j^k} \tag{6.9}$$

Thereby,

$$\left[X_i^k = \sum_j \left(a_{ij} X_j\right)^k \cdot \text{Prob}\left(x_{ij}^k\right) + x_i^k \cdot \text{Prob}\left(x_{ij}^k\right)\right] \approx \sum_j \left(a_{ij} X_j\right)^k \cdot \text{Prob}\left(x_{ij}^k\right), \tag{6.10}$$

This equation reduces to, $X_i^k = \sum_j b_{ij}^k \cdot X_j^k$; where, $b_{ij}^k = \left[\frac{a_{ij}}{\left(1-a_{ii}\right)^k}\right]$. (6.11)

The totality of these equations for $i = 1, 2, \ldots, n$ represents the circular causation relations between the variables.

The wellbeing objective criterion is denoted by,

$W(\theta) = W(\mathbf{X}(\boldsymbol{\theta}))$ in the conceptual abstraction stage of its explanation; $\mathbf{X}(\theta) = \{X_1, X_2, \ldots, X_n\}[\theta]$.

The conceptual form of $W(.)$ then takes the measurable quantitative form by the Implicit Function theorem of the continuous differentiable form,

$$\theta = F\left(\mathbf{X}\left(\theta\right)\right). \tag{6.12}$$

$\{\theta\}$ statistical values are computed corresponding to the $\mathbf{X}(\theta)$ values in accordance with the highest-ranked ordinal value being assigned to the lowest positive value of $x_i^k\left(\theta\right)$. Rest of the θ values are thereby prorated.[8] Such a ranking of θ values follows the following explanation:

Islamic instrument-specific savings, $x_i^k \to 0$, as mobilization of savings increases into real investment increases; debt financing is ruled out in a consequently transforming money, finance, and real economy (MFRE) interrelationship. Residual savings as bank savings decline. All residual savings other than these ones are already recycled into the real economy through foreign trade financing and resource mobilization.

The shari'ah-compliance approach being based on short-term 'aqd-rulings, and not having sound formulation of intertemporal rates, as in ijara-type financing, was shown above to be incapable of establishing any such approach to apply to a mix of various kinds of premiums. Consequently, no portfolio theory of organically unified Islamic financing instruments (securities) can be expected to be derived. As well, no behavioural expected preferences of propensities to risk and return in such a case of Islamic capital market transformation can be explained by shari'ah-compliance.

Islamic finance academics have thus submitted themselves to the usual mainstream theory of risk and return, which copies CAPM. Yet such an academic submission will now be shown to be incorrect for explaining investor behaviour in an organically interdependent Islamic securities market.

Risk and Return Diversification Theory in Tawhidi Perspective of Unity of Knowledge

The Taylor expansion of an analytic function that was explained in Footnote 3 depended upon the mainstream theory of diminishing marginal utility and maximization behaviour that leads to a non-learning world characterized by the end of novelty in learning (Shackle, 1972; Jeans, 1970). None

[8] Lowest non-zero $\left(x_i^k\right)^*$ value is assigned an ordinal statistical value of $\theta = 10$ (say; or any other highest-ranked integer). Remaining of the $\left\{x_i^k\right\}$ values assume their prorated values by the formula, $\left(x_i^k / x_j^{k*}\right)\cdot 10$. The generated statistical θ values side by side of the corresponding $X_i^k\left(\theta\right)$, now estimate the equation, $\theta = F(\mathbf{X}(\theta))$. Estimation is followed by simulation.

of these assumptions co-exists with the Tawhidi analytical methodology in respect of its precepts of interaction, integration, and evolutionary learning. The quadratic term that is negatively explained by the second-order derivative is the result of the above-mentioned neoclassical postulates. This same coefficient is not negative according to the Tawhidi evolutionary learning methodology. Thereby, the non-Cartesian domain in the learning flux (i.e. not on Cartesian surface, hence in a topological space) represents an amorphous body of complementary relational forms driven by unity of knowledge defining the continuous flood of evolutionary tuples.

We write for this analytical property,

$$\left(\frac{d}{d\theta}\right)\left[\overset{\text{interaction}}{\cup}\ \overset{\text{integration}}{\cap}\left\{E\left(\mathbf{X}(\theta)\right),\operatorname{cov}\left(\mathbf{X}(\theta)\right),t(\theta)\right\}\right]>0. \tag{6.13}$$

Every point on the amorphous evolutionary domain is like the evolutionary point denoted earlier by $h(\theta)$ in Fig. 6.2, floating in knowledge, space, and time dimensions.

Now the differentiating explanation between the Tawhidi attitude to risk and return and the conventional finance theory is this: The heightened propensity to assume risk to earn higher financial return in Tawhidi methodological worldview implies that interest rate ('*i*') is not the substitution price of rates of return, '*r*' (i.e. financial substitution defined by i/r is inadmissible). This property would incite investors with risk-aversion behaviour to allocate wealth into interest bearing risk-free (or constant risk) portfolio.

There is also the explanation in respect of the Tawhidi explained risk-return behaviour of the type explained here. This involves portfolio diversification in both systematic and idiosyncratic risks. The Qur'an and sunnah promote the market orientation of the economy. Islamic banks ought to generate their rates of return from market exchange by mobilizing savings as financial resources into blissful real resources. This economic function causes production diversification and increased shareholding.

Thereby,

$$\frac{\text{total risk capital}\left(\text{market risk}\right)}{\left(\text{production diversification}\times\text{number of shareholders},\text{sh}_s\right)} =\left(\frac{\operatorname{Var}\left(\mathbf{X}(\theta)\right)}{\left[\prod_s T_s\left(q_s^{\alpha}\left(X(\theta)\right)\times q_s^{\alpha}\left(X(\theta)\right)\times \text{sh}_s(\theta)\right]}\right). \tag{6.14}$$

This formula gives the unit risk of a venture. The compounded factors of unit risk signify the degree of effective risk-diversification possible. In this formula of unit risk, the technological factor denoted by $T_s\left(q_s^{\alpha}\left(\mathbf{X}(\theta)\right)\right)$ can be included. Thereby, even as risk increases in the amorphous flood of $h(\theta)$-like events, it is the effective risk-diversification, as defined above, that influences the behaviour of the investor in respect of the complementary propensity towards risk and return.

As the number of shareholders expands to take account of stakeholders (sh), the risk base (Var($\mathbf{X}(\theta)$)) and the diversified products $\prod_s q_s^{\alpha s}$ expand over social and spiritual goods (Zohar & Marshall, 2004), then the unit cost of risk reduces by holding up ethical transformation of the economy. This consequence shows up by complementarities between the risk, return, production, financing, stakeholding, and rates of return contrary to interest rate and so on in the wellbeing objective criterion. The inter-variable complementarities explained in and by the quantitative form of the wellbeing function and the circular causation relations, respectively, put into consilience the Tawhidi episteme of unity of knowledge (Wilson, 1998) in the financial world-system, economy, and society. All these conscious financial behaviours yield results of substantive nature and are quite the contrary to mainstream financial economics that Islamic academics and practitioners have imitated.

Pervasive Problem of Non-sustainability and Distortionary Allocation of Capital in the Long Run in Shari'ah-Compliance

Many other Islamic financing rulings under the shari'ah-compliance approach are based on fiqh (jurisprudence) and fatawa (theological ruling) that arise from a closed doctrinaire and institutions (e.g. AAOIFI, OIC Fiqh Council). Yet in respect of analytical argument such rulings can be grossly flawed. We consider here, firstly, the ruling given in favour of what is termed as 'profit equalization'. The ruling is to compensate for lower performance of rates of profits in unprofitable enterprises by distributing a portion from the reserve fund that would bring up the enterprise profit rate to a level of acceptable rate of profitability. It is felt that such an enhanced rate would then establish the upfront performance semblance of enterprises and investors and thus show them to be more competitive and profitable.

Profit Equalization

The economic argument against profit equalization method is that such a financing is like a mark-up of murabaha that has no real asset backing for determining this rate, except that the mark-up is based on a mutually agreed-upon subjective probability of occurrence of contingencies associated with a low performance of hire-purchase of an asset.

In the profit equalization plan set on a reserve fund, the guarantee of a desired rate is equivalent to the unproductive nature of interest rate on idle bank savings for the time that it is held up at the bank. If this practice is allowed to continue, then the portfolio of contributors and their amount of contributions into the profit equalization fund will distort the legitimacy of individual productive claims on the basis of earned profit rates. A distortionary effect of redistribution of available resources ensues (Nozick, 1974).

The way to escape such distortionary consequences would be to utilize the inter-security (instrument) input-output portfolio relationship that was explained earlier. Thereby, market-determined rates can be generated to replace undue and exogenously determined profit mark-up. Finally, in the macroeconomic sense, the extra quantity of money supplied via unproductive means shows that the rate of growth of money supply is higher than the rate of growth of productivity denoted by the rate of profit. The excess supply of money now becomes a precursor to expected inflationary pressure. Likewise, we find in the precautionary demand for money that an interest portion of the demand is upheld to fill up the gap in potential savings to meet economic adversity.

Tawarruq Financing

A worse type of speculative pricing is found with tawarruq, whereby the Islamic financial institution buys an asset from a dealer at an agreed-upon deferred price that the financial firm will pay to the dealer on maturity of the contract. The financial institution then sells the asset to another buyer at a cash price to raise cash value from sale. Thereby, the difference between the deferred price and cash value is clearly a speculative profit differential in favour of the financial enterprise at the detriment of the first seller, who could otherwise have sold the items directly to the buyer in the absence of a mark-up without the loading of profit margin and a non-market determined ijara rate that can be interpreted as riba. Such was the basis of interest in the theory of roundaboutness of production instead of direct dealing, as explained by the Austrian School of Economics (von-Bawerk, 1890).

The Sukuk Problem of Shari'ah-Compliance

Consider the following financing of a trade in material: Client B needs material that A has. The SPV purchases the material from A on deferred payment basis and supplies it to B. If B cannot pay, then SPV raises the returns of the sale to B from sukuk holders or holders of bonds sold by the SPV to raise funds for purchase of materials required by partners. At the end of the maturity date of the sukuk, SPV will participate with A and B on the returns from the operations of the sukuk. This example gives rise to multiple trade. Yet what is the problem with it? The multiple sukuk trade is based on debt instruments in which the deferred financing will involve risk and uncertainty to poise the partners to financial adversity against the fixity to pay the deferred price of the material. The reduction of unit risk for the gains to partners is ignored, in the way this principle of risk-diversification was explained earlier.

Sukuk and Takaful Relationship in Interest-Free Financing Situation

To avoid such episodes, financing and trade both ought to be based on debt-free transactions even though on a participatory basis. Debt-free instruments will arise from cash payments on re-takaful basis at SPV to pay for the needed merchandise to A received from B. The principal is thus paid off. On the matter of profit-sharing from the real sector operation of the sukuk fund, complementary profit-sharing participation can be undertaken.

Re-takaful when combined with other participating financing partners can be an effective avenue towards lowering of unit risk, on the basis of which investors develop propensity to undertake risk and return with positive complementarities between them. We now have a 3 × 3 input-output matrix in inter-instrument cash flows for trade in goods and services. In this arrangement, risk-diversification, production-diversification, and increase in number of stakeholders are realized together.

This is an example of sukuk with murabaha contract. A rent with mark-up can be divided into a market—actualized profit margin at every point of time. The residual mark-up on rent can be seen as an investment to generate a profit share between the rentier and the financing company. The resulting effect of such financing is felt in its contribution to wellbeing objective via disaggregation of sukuk at the small enterprise level. An example is the sukuk financing of low-cost housing project as social capital.

Conclusion

The shari'ah-compliance literature comprises a cluttered body of complicated reading. Its heavy reliance on Arabic nomenclature makes unpleasant vocabulary and roundabout explanation. The study of Shari'ah Advisory Council report unfolds unnecessary matters of simplicity that common sense can readily establish in the wellbeing perspective of Islamic financing, sale, and ownership.

The origin of shari'ah-compliance is based on institutional fatawas (religious ruling) and weakly narrated ahadith (prophetic sayings), never heard of. On the other hand, the indispensable altar of the Tawhidi methodological episteme of unity of knowledge and the details of the world-system is nowhere to be found in shari'ah-compliance approach. The experts of shari'ah-compliance field have not acquired this know-how. Shari'ah-compliance is not intellectually ready to deliberate on the epistemological and formal perspectives as of Tawhidi methodological worldview.

The Tawhidi methodological worldview of consilience meaning monotheistic unity of knowledge is premised on simple but rich epistemological grounds. The Tawhidi methodology is derived from the rule of organic relational unity of being and becoming between all good things of life that the Qur'an and the sunnah have not forbidden or have recommended.

Within this derivation and setting up of the Tawhidi law of unity of knowledge in respect of the generality and details of all matters, two approaches of ethico-economic choice are used. Firstly, there is the algorithmic approach to find out inter-variable complementarities as unity of relations between the good things and actions as the Qur'an and the sunnah recommend or do not forbid. Secondly, the algorithmic results are examined by discourse among participating agents to authenticate or revise the algorithmic results on inter-variable circular causation relations by inter-causal complementarities between the selected variables.

Thus, the Tawhidi methodology of legitimacy of ethico-economic choices would negate some of the rules derived by shari'ah-compliance. The rest of the rules will be subject to Tawhidi investigation, while avoiding the unnecessary simplicities as otherwise manifest by verbosity of rulings by shari'ah-compliance.

In conclusion, we note the following specific case that establishes our criticism here. Consider the following case. Shari'ah Advisory Council writes on the interest of the insured in taking out takaful (insurance): "If a person applied for a life insurance policy or a family takaful plan for another person, and nominated himself as the beneficiary, his application will not be approved because there is no insurable interest." This is an inaccurate statement of a simple rule that is automatically valid in the event of a contingency happening. It is a simple matter, hardly requiring substantive shari'ah-compliance attention to know that there exists conditional probability of survival and usufruct in the event of death of the insured. For, on the death of the second party the first party inherits the insurance benefits to perform the functions of the originally insured. The same is true of multiple partners upon survival. Usufruct rules of caring applies here at any point during the life time of individual survival and participatory business. The same takaful rule would exist in the case of all different kinds of contingencies. Indeed, a participatory form of insurance is always the rule of the partnership security.

In the case of the Tawhidi worldview of unity of knowledge, the contingent insurance based on partnership is a recommended matter. Its functioning shows the organic unity of relations that exists in the insurance spirit for sustaining wellbeing in participated partnership. The acceptability to insurance of the resultant type does not require unnecessary legitimation. In fact, the contingency interest in takaful is essentially legitimate under the Tawhidi principle of epistemic unity of knowledge as complementarity of inter-causal variables of the organically unified wellbeing.

All debt instruments are unnecessary and are contrary to the spirit of avoiding shadow rates of interest in deferred financial transactions. In this chapter, it was explained how the Tawhidi methodological objective of wellbeing, which is evaluated subject to circular causation, does away with debt instrument financing in an interactively integrated dynamic version of the key debt-retiring MFRE model.

On a generalized basis of rejecting this central nature of pricing and valuation by deferment in shari'ah-compliance approach, and its replacement by the Tawhidi methodological worldview of unity of knowledge, we formalize replacement of all debt instruments that are presently being legitimated under shari'ah-compliance. This would also result in formalization of rate-setting formulas in valuation of contingencies.

We explain the Tawhidi opposite case of valuation in Fig. 6.3. The practice of 'aqd contract under shari'ah-compliance is replaced by the interrelational model of wellbeing that explains the organically unified inter-variable causality over the knowledge, space, and time dimensions. It applies to all forms of valuations. The last words are that Tawhid is the ultimate ontological law of 'everything'. The maqasid al-shari'ah is a gateway to understand the ontological law of Tawhid and all world-systems,

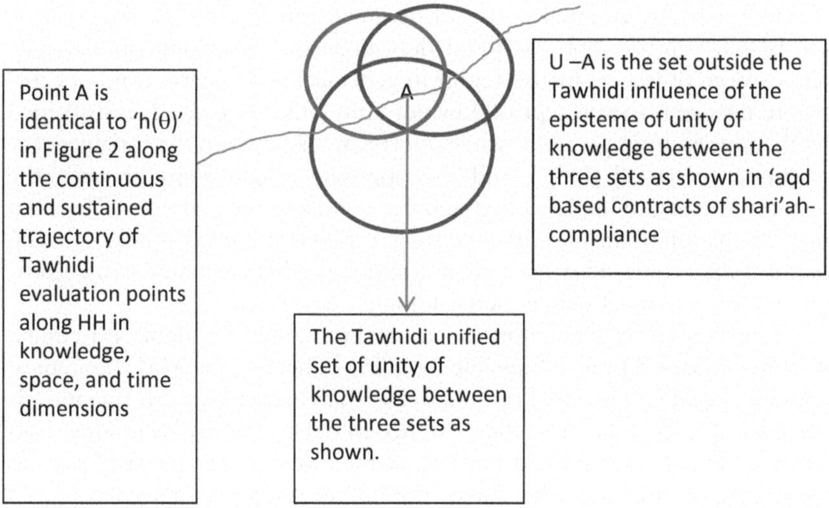

Fig. 6.3 Tawhidi evaluation worldview of sustainability contrary to 'aqd-based shari'ah-compliance

though it has not yet proved itself in this capacity (Choudhury, 2015a, op cit). The shari'ah-compliance area is a distortion of the actual spirit of Tawhid and maqasid al-shari'ah.

Appendix: Islamic Financial Structure of the Firm in a Mix of Debt, Equity, and Tax

The two papers, namely by Dzolkarnaini and Minhat (2016) and Singh and Sheng (2012), unpublished though, could not answer the question of irrelevance of the Modigliani and Miller (MM) theorem in terms of Islamic economics and finance. The authors held up their unquestioned belief on the acceptability of debt in Islamic financing. This chapter argues that debt is the grand mistake and that the so-called Islamic economists and shari'ah-compliance have continued to legitimize. For, to make the MM theorem regarding the optimal debt–equity ratio as the basis of irrelevance of debt and equity in the optimal financing structure of a corporation requires fundamentally the use of the neoclassical assumptions of economic rationality. This comprises the postulates of full information, perfect capital market, zero transaction cost, and the existence of resource allocation conditions to establish optimal valuation of shareholders' wealth.

These starting axiomatic bases of neoclassicism destroy the relevance of the Islamic venture capital without debt financing. The foundational epistemic axiom of Islamic methodology in generality and in particularity of corporate finance is the precept of Tawhidi unity of knowledge. It is explained by the Qur'an in respect of organic interrelations between the variables representing the good things; and the opposing relationships between the unrecommended things and between the good and the bad ones. This kind of confirmation of the nature of interrelations between choices is the starting point of the evolutionary learning methodology that emanates across learning processes in the Tawhidi methodological worldview.

According to shari'ah-compliance on the question of debt and equity, debt is considered to be admissible for a firm. Thereby, the MM theorem of debt and equity choice that optimizes the shareholder's wealth valuation is considered to be valid. Yet, the existence of debt gives rise to interest paid on debt. The debt-cancelling interest payment in turn can be written off as tax deduction. On the other hand, the higher is equity participation, and thereby the lower is the debt for financing shareholder's optimal wealth, the higher is expected to be the taxation on dividends arising from equity investment. At the end, the mix of debt–equity structure of the corporation establishes a trade-off between debt and equity in such a manner that debt can never be avoided. Thereby, interest rate on debt cannot be avoided.

In such Modigliani and Miller's treatment of the theory of debt–equity composition in the optimal structure of corporation capital investment, there is no process of evolutionary learning relating to the phasing out of interest, as this would manifest in Tawhidi methodology. Debt and taxation must exist perpetually in the MM theorem and its imitation in Islamic finance. In fact, the Islamic finance field is now at its juncture of worsening thought and application regarding the moral and ethical objectivity of the wellbeing criterion of Tawhidi unity of knowledge. Such remiss holds back the understanding of the debt problem in Islamic socio-economic transformation.

Singh and Sheng (op cit) rightly analysed that, in the final goal of the Islamic financial order regarding converting all financial capital into equity, the optimal debt–equity (in this order) ratio will be zero. Now with zero rate of interest and all capital invested in equity (thus interest rate replaced by rate of return), the final point of allocation of capital shows a static corner point without learning (Fig. 6.4). The debt–equity line is then the whole of the horizontal axis. This causes a contradiction between the neoclassical economic assumptions of optimality and steady-state equilibrium when the learning point is caused by the Tawhidi methodological

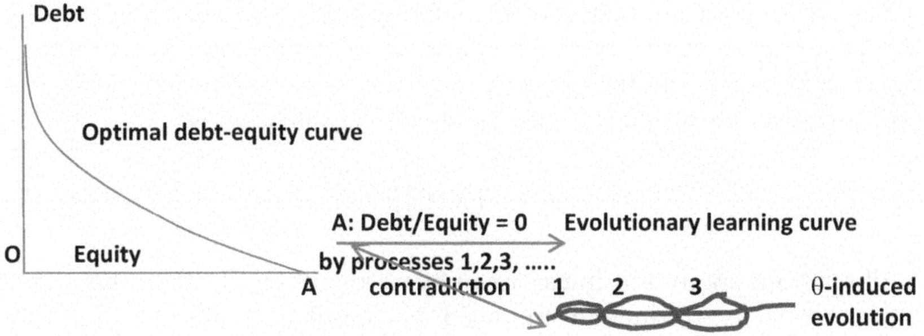

Fig. 6.4 Tawhidi evolutionary learning phenomenon versus MM theorem of debt–equity optimality of capital structure

worldview all along the horizontal axis. Such an evolutionary learning phenomenon cannot be shown along the horizontal line in neoclassical economics and its presence in Islamic economics and finance. See Fig. 6.4.

The Tawhidi methodological functioning of resource allocation by the wellbeing objective criterion (also pointed out by Singh and Sheng for a social transformation into the Islamic capital market), provides the following formalism between the variables: debt (D), equity (E), interest rate (i), taxes (T), shareholder's valuation of wealth (V), dividends Dv, and the learning parameter along the trajectory of Islamic transformation (θ).

By our earlier definition of the wellbeing function,

$$W(\theta) = W\left(\Delta_\theta(E/D), \Delta_\theta(Dv/i), \Delta_\theta(E/T), \Delta_\theta(Dv/V), t\right)[\theta], \tag{6.15}$$

to evaluate the degree of positive complementarities between the variables as shown here.

That is,

$$\begin{aligned} \mathbf{X}(\theta) = \{&\Delta_\theta(E/D)\uparrow(\downarrow) \leftrightarrow \Delta_\theta(Dv/i)\uparrow(\downarrow) \\ &\leftrightarrow \Delta_\theta(E/T)\uparrow(\downarrow) \leftrightarrow \Delta_\theta(Dv/V)\uparrow(\downarrow)\}; \end{aligned} \tag{6.16}$$

over time. $\mathbf{X}(\theta)$ is the vector of variables.

This string of relations explains circular causation relations between the variables in terms of the actual, 'as is' (estimation) followed by 'as it ought to be' (simulation) states of inter-variable relations. By doing so, there is an algorithmic reconstruction of the coefficients in the wellbeing and the circular causation model system towards transforming into 'as it ought to be' states by inter-variable strong complementarities. Now the quantitative expression for wellbeing used for empirical valuation is,

$$\theta = F(X(\theta)). \tag{6.17}$$

All relations are in non-linear form, by virtue of the fact that, the evaluated coefficients (estimated followed by simulated) are continuously changeable by choice, and are thereby, dynamic in form. The decreasing case shown by (↓) is the mainstream one. This state of valuation does not reflect the Tawhidi episteme concerning the knowledge-induced variables.

In this case, the MM theorem replaces $\{\theta\}$ by $\{t\}$. No predictable movements of the signs of the coefficients can establish the inter-variable circular causation relations inherent in the series of equations arising from

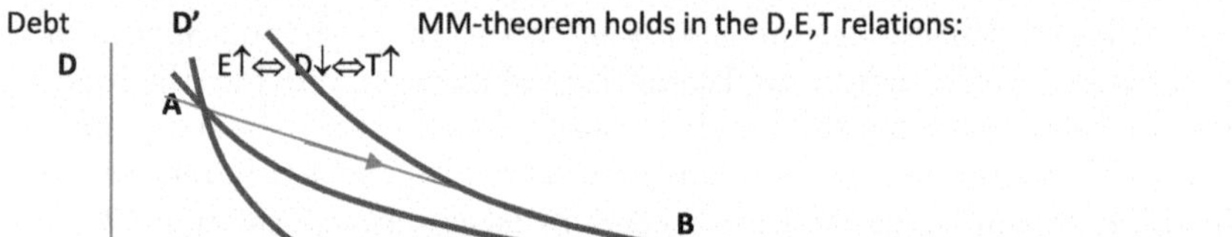

Fig. 6.5 θ-induced Tawhidi versus MM theorem relating to debt, equity, and tax relations in the capital structure

expressions (6.16) and (6.17). Therefore, policy inferences cannot be formed, as otherwise would be required for transformation into the true nature of Islamic economy. In this case, the pervasively positive complementarities between the variables in the 'as it ought to be' simulated scenario marks the inferences concerning true Islamic transformation.

For instance, in a transforming Islamic economy of the life-fulfilment regime of development and sustainability, $\Delta_\theta T(\theta)\downarrow$, as $\theta\uparrow$. Figure 6.5 explains the contrary implications between the Islamic case and the MM theorem connecting the variables mentioned in expression (6.16) in mainstream economic theory and in shari'ah-compliance approach with debt financing.

References

Asad, M. (1987). *This Law of Ours* (p. 23). Gibraltar: Dar al-Andalus.

Auda, J. (2008). *Maqasid al-Shari'ah as Philosophy of Islamic Law, a Systems Approach*. London: The International Institute of Islamic Thought.

Bruteau, B. (1997). *God's Ecstasy* (p. 179). New York, NY: The Crossroad Publishing Co.

Buchman, D. (1998). *The Niche of Lights* (p. xiii). Provo, Utah: Brigham Young University Press.

Choudhury, M. A. (2011). Overlapping Generation Model for Islamic Asset Valuation: A Phenomenological Application. Chapter 8 of his, *Islamic Economics and Finance, an Epistemological Inquiry*, (Contributions to Economic Analysis), Emerald Group Publication, Bingley, UK.

Choudhury, M. A. (2015a). Res extensa et res cogitans demaqasid as-shari'ah. *International Journal of Law and Management, 57*(6).

Choudhury, M. A. (2015b). Subjective Probability in Financial Valuation: Contrasting Meanings. *Journal of Financial Research and Accounting, 13*, 1.

Choudhury, M. A. (2016). Monetary and Fiscal Spending, Complementarities to Attain Socioeconomic Sustainability. *ACRN Journal of Finance and Risk Perspectives, 4*, 3.

Coulson, N. J. (1984). *Commercial Law in the Gulf States: The Islamic Legal Tradition*. London: Graham & Trotman.

Dzolkarnaini, N., & Minhat, M. (2016). In Search of a Theory of Corporate Financing and Islamic Financial Instruments. Paper presented at the 8th International Conference in Islamic Economics and Finance, International Islamic University Malaysia, Kuala Lumpur, Malaysia, Unpublished.

Gel'fand, I. M. (1961). *Lectures on Linear Algebra*. New York, NY: Interscience Publishers, Inc.

Hassan, H. (2002). Contracts in Islamic Law: The Principles of Commutative Justice and Liberality. *Journal of Islamic Studies, 13*(3), 257–297.

Hogg, R. V., & Craig, A. T. (1965). *Introduction to Mathematical Statistics.* New York, NY: Macmillan Co.
Jeans, W. H. (1970). Borrowing and Capital Structure. Chapter 6 of his, *The Analytical Theory of Finance.* New York: Holt, Rinehart and Winston, Inc..
Kelley, C. T. (2003). Solving Nonlinear Equations with Newton's Method. No 1 in *Fundamentals of Algorithms,* SIAM, 2003. ISBN 0-89871-546-6.
Lawson, T., & Pesaran, H. (1985). Methodological Issues in Keynes' Economics: An Introduction. In *their Keynes' Economics, Methodological Issues* (pp. 1–9). London: Routledge.
Misri, R. (1987). *Masraf al-Tanmiyah al-Islami aw MuhawalahJadidah fi al-Rib-awa al-Faidahwa al-Bank (Essay on Integration of a Development Bank in an Islamic Society: The Problems of the Islamic Conception of Interest).* Beirut: Muassasah al-Risalah.
Nozick, R. (1974). Distributive Justice. *Philosophy and Public Affairs, 3,* 45–126.
O'Donnell, R. M. (1989). Some Philosophical Background. In *Keynes: Philosophy, Economics & Politics* (pp. 11–28). London: Macmillan.
Robinson, J. (1961). Prelude to a Critique of Economic Theory. *Oxford Economic Papers, 13,* 7–14.
Shackle, G. L. S. (1972). *Epistemics and Economics.* Cambridge: Cambridge University Press.
Shari'ah Advisory Council. (2016). *Report.* Kuala Lumpur, Malaysia.
Singh, A., & Sheng, A. (2012). Islamic Finance Revisited: Conceptual and Analytical Issues from the Perspective of Conventional Economics. Online at https://mpra.ub.uni-muenchen.de/39007/. MPRA Paper No. 39007, posted 26 May 2012, visited 28 December 2016.
Von-Bawerk, E. B. (1890). *Capital and Interest.* London: William Smart.
Weeramantry, C. G. (2001). *Islamic Jurisprudence, an International Perspective.* Kuala Lumpur: The Other Press.
Wilson, E. O. (1998). *Consilience, Unity of Knowledge.* New York: Vantage Press.
Zohar, D., & Marshall, I. (2004). *Spiritual Capital.* San Francisco, CA: Berrett-Koehler Publishers, Inc.

CHAPTER 7

A Critical Examination of the Shari'ah in the Light of the Legal Commercial Report of ISRA

Imposed Meanings in Shari'ah by Government Authority

Such enforced conditions following the divergent interpretation of rules other than those prescribed overtly by the Qur'an ought not to be imposed by state institutions as in Malaysia, and the royal decrees as in Saudi Arabia. Instead, a learning process in shari'ah ought to be engendered by participatory discourse on specific matters by the conscious public at large. For this learning process to take effect, the conscious public ought to be adduced to uphold the most social, ethical, and moral principle of unity of knowledge in evaluating wellbeing in and across 'everything'. Through such confirming social practices towards upholding the Tawhidi law of unity of knowledge, the moral and ethical law for common wellbeing would be better accepted by society at large. Such rules are best adaptable to socio-scientific cases. It is then possible to understand Tawhid as the law governing the heavens and the earth and of all that is below the earth and above the earth. In this regard, the Qur'an (20:6) declares: "Unto Him belongs all that is in the heavens and all that is on earth, as well as all that is between them and all that is beneath the soil." Contrary to such guidance of the Qur'an, the Muslim world ever after the demise of the Prophet Muhammad and the pious caliphs proved to be autocratic in its rulers.

M. A. Choudhury, *Tawhid and Shari'ah*,
https://doi.org/10.1007/978-3-030-49087-4_7

The Vista of Socio-scientific Evolutionary Learning in Unity of Knowledge

The shari'ah, mostly as a socio-scientific reasoning, emerged as an obtrusive law of the courts under such decadent rulers in contradistinction of the origin of maqasid al-shari'ah at the time of the learned Islamic scholars of classical Islamic age. It is this area of revisionism of Islamic law as an intellectual study that is now subject to social intervention in the law of the universal and unique law that is conveyed by the primal ontological origins of the Qur'an and sunnah. This is Tawhid as the primal law of 'everything'. It is nonetheless maintained and sustained over the dimensions of knowledge, space, and time by its methodological worldview of unity of knowledge. The moral, ethical, and social elements are thus induced into worldly affairs of science integrated with social issues along the same foundation of monotheistic unity of knowledge. It is interesting to note that this is also what modern socio-scientific thought has always pursued. Kant sought such a transcendental holism by his writings on consciousness.

Kant wrote:

> Two things fill the mind with ever new and increasing awe and admiration the more frequently and continuously reflection is occupied with them; the starred heaven above me and the moral law within me. I ought not to seek either outside my field of vision, as though they were either shrouded in obscurity or were visionary. I see them confronting me and link them immediately with the consciousness of my existence.

Then there is the ponderable extract from Wilson: "Looked at in proper perspective, God subsumes science, science does not subsume God. … Scientific research is not designed to explore all of the wondrous varieties of human experience. The idea of God in contrast, has the capacity to explain 'everything', not just measurable phenomena, but phenomena personally felt and sublimely sensed, including Revelation that can be communicated solely through spiritual channels."

Foucault promoted for the establishment of a discursive society in the following words: Episteme is defined by Foucault in the following words; we accept this definition as our understanding of the totality of functional relationship between knowledge, cognition (res cogitans), and world-system (res extensa): "By *episteme* we mean ... the total set of relations that unite, at a given period, the discursive practices that give rise to epistemological figures, sciences, and possibly formalized systems ... The episteme is not a form of knowledge (*connaissance*) or type of rationality which, crossing the boundaries of the most varied sciences, manifests the sovereign unity of a subject, a spirit, or a period; it is the totality of relations that can be discovered, for a given period, between the sciences when one analyses them at the level of discursive regularities."

Is the quest for socio-scientific res extensa and res cogitans deprived of Tawhidi socio-scientific unity of being and becoming? Contrarily, Tawhid establishes the foundational reasoning of the universal law. The Qur'an (87:18,19) declares in this regard of Tawhid as the most primordial abiding law: "And this is in the Books of the earliest (Revelation) (87:18); the Books of Abraham and Moses (87:19)."

Qur'anic Meaning of Tawhid as the Law and Shari'ah as the Right Way of Tawhid

According to the Qur'an and thereby emulated by the sunnah, the meaning of shari'ah is the right way, not the law. Such a unique and universal right way was bestowed on all the messengers of Allah and thus all communities over the full time of divine revelation. When understood in this way, the shari'ah way is equivalent to the revealed way of Tawhid as unity of the divine law over universal history. Such an interpretation can be derived from the verses (Qur'an 42:13; 42:15, 4:18):

> He has prescribed for you the Way [Shara'alakumminad-Diin] which He has enjoined upon Noah, and which We have revealed to you, and which We had enjoined upon Abraham, Moses and Jesus, declaring, 'Maintain the Way, and be not divided in it.'... (42:13)

> So summon to this [unity of religion], and be steadfast, just as you have been commanded, and do not follow their desires, and say, 'I believe in whatever Book God has sent down. I have been commanded to be just among you. God is our Lord and your Lord. Our deeds belong to us and your deeds belong to you. There is no argument between us and you. God will bring us together and toward Him is the destination.' (42: 15)

> Then We set you upon the right Way [Shari'ahtim-minal- 'amri]; so follow it, and do not follow the desires of those who do not know. (45: 18)

Thus as THE LAW that overarches 'everything' in mind, matter, evidence and the organic holism, shari'ah is superseded by Tawhidi worldview of unity of being and becoming concerning the relationship of divine law with the world-system. The shari'ah contradiction to Tawhid as the Qur'anic law in terms of its methodological worldview of organic unity of knowledge (Qur'an 36:36) can be explained by the failure of organismic relations of unity of knowledge in Fig. 7.1.

In Fig. 7.1, the above-mentioned Qur'anic verses are contradicted along the discontinuous paths shown by the broken paths shown by 'XXX'. Such are the illogical paths of shari'ah derivation both for maqasid al-shari'ah and 'shari'ah-compliance' in present-day usage. Thereby, the Qur'anic meaning of maqasid al-shari'ah as the Right Way derived from Tawhid as law is disabled in the face of its present days' humanly concocted interpretation differently as the Right Way in the light of Tawhid as the law. The maqasid al-shari'ah and thereby all of shari'ah as the right way arising from the primordial ontology of Tawhid as law can only be established along the uninterrupted straight path from the Beginning to the End as shown in Fig. 7.1. The straight path of perfect truth and divine

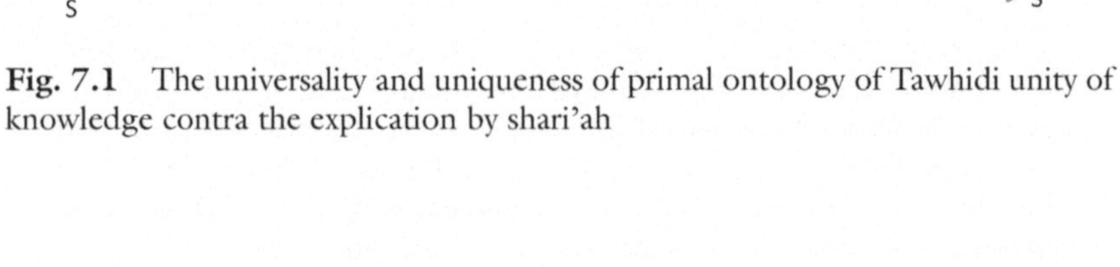

Fig. 7.1 The universality and uniqueness of primal ontology of Tawhidi unity of knowledge contra the explication by shari'ah

blessing is called siratul mustaqim, after giving allowance for the errors and recovery of perturbations in evolutionary learning to seek evolutionary equilibrium (Grandmont, 1989) around the worldly inaccessibility of straight path, leaving this ultimate phenomenon be attained in the Akhira.[1]

Abdullah Yusuf Ali writes (footnote of Qur'an translation, 4756): "*Shariah* is best translated the 'Right Way of Religion', which is wider than the mere formal rites and legal provisions, which mostly came in the Medina period, long after this Meccan verse had been revealed." (Abdullah Yusuf Ali, *The Holy Qur'an: Text, Translation and Commentary*, New York: TahrikeTarsile Qur'an, Footnote no. 4756). In combining the above-mentioned verses on the nature and meaning of shari'ah, the 'Right Way of Religion' is the way emanating from and returning to the continuity of Tawhid across the conscious expanse in continuum. Such is the phenomenology of Tawhid as the Law, not of any mistaken understanding of referring to shari'ah contrary to this primal ontological leaning of Tawhid as law and its organic interrelationship with 'everything'. Figure 7.1 establishes this essential nature of Tawhid contra shari'ah.

Distortion of Shari'ah as the Right Way in the Light of Tawhid

Every noteworthy shari'ah study in the literature (Shari'ah Advisory Council) has ignored Tawhid as law. The literature and their scholars and sects have failed to undertake a bold criticism of shari'ah in all its forms on grounds of the organic unity of knowledge that Tawhid bestows as socio-scientific methodological worldview by its universality and uniqueness properties concerning 'everything'. A perusal of Islamic Commercial Law Report by ISRA and Thomson Reuter points out a cursory write-up on maqasid al-shari'ah without any basis of its legitimacy in the light of Tawhid as the ontological law of organic unity of knowledge. The reason is unknown as to why this negligence is allowed to prevail, and why alternative scholarly viewpoints are not examined. Yet the limitation of the approach appears to reject the existing Tawhidi literature in the area of

[1] Qur'an (52:52,53): "And thus We have revealed to you an inspiration of Our command. You did not know what is the Book or [what is] faith, but We have made it a light by which We guide whom We will of Our servants. And indeed, [O Muhammad], you guide to a straight path (52:52)—The path of Allah, to whom belongs whatever is in the heavens and whatever is on the earth. Unquestionably, to Allah do [all] matters evolve (52:53)."

transdisciplinary coverage of Islamic economics, finance, science, and society that truly forms the field of an integrated socio-scientific study.

Consequent to the ambivalence of any critical study of shari'ah contrasting with Tawhid as the ultimate conceptual and applied law, certain major relevance of Islamic financing has been overlooked. The result thereby is that, as described in the ISRA-Thomson publication, the objective, purpose, modes, and selection of financing fail to qualify as 'Islamic-financing'. Firstly, upon what does the maqasid al-shari'ah rest and arise from for a sustainable continuity of Islamic financing? Here the ontological methodology of unity of knowledge of Tawhid and $t =$ its interrelationship with the world-system remains unknown, A mere statement that maqasid al-shari'ah originates in the Qur'an, and sunnah is an ignorant way of pointing out the roots of academia on the matter of Tawhidi reference.

The above-mentioned kind of ambivalence concerning Tawhidi methodological worldview and shari'ah in all its forms has resulted in choices that clamour the mere announcement of an ever-growing capital market of the so-called shari'ah product development and capturing shareholders' penetration of the capital market. The resulting perception to the wider field of moral, ethical, and social valuation of assets in the poverty-stricken Muslim world with all its multidimensional poverty-generating factors has been subdued by the repeated show of percentages of market shares. This is a sorry state of business mind, and cannot speak for any kind of matching with corporate social responsibility (CSR) and Shari'ah Index (SRI). The shareholder model as opposed to stakeholder model is still the yard stick of measuring success of 'Islamic' financial assets and expansion.

The adverse results on financial valuation continue as a result of upholding the juristic, maqasid, and impersonal attributes contrary to the true Tawhidi worldview arising from the Qur'an and sunnah. Among these many logical and moral/ethical failures of financial, market, and institutional evaluations, some are pointed out here:

1. In the preoccupation of evaluating Islamic finance by its capital market penetration for shareholders profitability, the ISRA-Thomson Commercial Report has forgotten the potential of academic contribution that research and erudition can make through market processes, institutions, and erudition. Most important in this direction of contribution would be to cultivate the moral-material values of method-

ological ways of socio-scientific integration and its empirical evaluation followed by inferences, choices, and policies.

2. The quantitative measure of maslaha index is flawed. It weighs choices and outcomes by institutionally imparted weights as exogenously assigned quantitative measures. Contrarily the Tawhidi approach to formulation and quantification of the wellbeing objective and purpose (estimation and simulation for reconstruction of the maslaha index) is an endogenous formalism with the data arising; hence weights arising from formal treatment.
3. The overwhelming governance by institutions as the government-appointed ones in the Muslim world for a long time concerning enforcement of shari'ah rules (ahkam) brings an external (exogenous) regulation of ideas. The common and independent participation, such as industrial participation, is overlooked. Thereby, socio-scientific development is suppressed by government-preferred will rather than ensuing a participatory grassroots direction of socio-scientific development.
4. The shareholder focus of capital asset and investment models in the portfolio of 'Islamic' finance shows a growing reliance on sukuk. Problems arise from the fact that the sukuk shareholders formal approach does not ameliorate the small-scale projects to enable the microenterprises and poverty alleviation, particularly in and through the rural economy. The end then is ignorance of the approach to life-fulfilment socio-scientific development, which otherwise is the outlook of necessaries (dururiyath), comforts (hajiyath), and refinements (tahsaniyath). All these categories of goods, services, and spiritual capital (Hansell, 2006) form the life fulfilment approach to development with the widening complex of complementary (participatory) interrelations. This latter perspective is vouched by the Tawhidi methodological worldview of monotheistic unity of knowledge and its induction of the world-system[2]
5. The overwhelming problem with sukuk to be categorized as 'Islamic' financing mode is that it revolves around debt instruments, of which almost all the so-called financing instruments, except profit- and loss sharing, mudarabah and musharakah, are debt instruments. The Qur'anic law on avoidance of financial interest (riba) is never avoided

[2] Hansell, G. (Aug. 6, 2006 essay). "What is spiritual capital? economics, religion, and conference 2006", Metanexus, online.

in the case of intertemporal valuation of assets by any of the 'Islamic' debit instruments. In the intertemporal case of asset valuation, the riba concept is different in terms of lateral addition of intra-systemic principle of possession (qabd) and the inter-systemic concept of valuation as a participatory project learns over knowledge, space, and time dimensions. This kind of valuation framework is formalized in the Tawhidi methodological framework of intra-systemic and inter-systemic knowledge-induced evolution. There is no equivalent method in the shari'ah approach to valuation. Thereby, the contractual ideas of qabd, trade, market exchange, and riba are flawed in the shari'ah approach of lateral aggregation. The differentiating principle is brought out in Fig. 7.2 in terms of intertemporal valuation across intra-systems combined with inter-system evolutionary learning. This is the methodological valuation results that is derived from the Tawhidi methodological worldview. It is not explainable by shari'ah.

6. The inadmissibility of the debt instruments that abound under the shari'ah humanly concocted guesses must be removed by the formal logic of unity of knowledge. This latter basis of Islamic financing across diversity of permissible choices is the unique approach that arises from the Tawhidi methodological worldview of unity of knowledge between the good things of life while avoiding the illicit ones. The underlying principle of extensive complementarities (participation) is the sure sign of upholding the Tawhidi organic unity of knowledge between the good things and constructing the corresponding basket of acceptable goods, services, and financing instruments.
7. The formal acceptance of the wellbeing function as the goal of maqasid al-shari'ah in the light of Tawhid as law is thus derived. The trajectory of the straight-path (siratul mustaqim), SS in Fig. 7.1, is thus

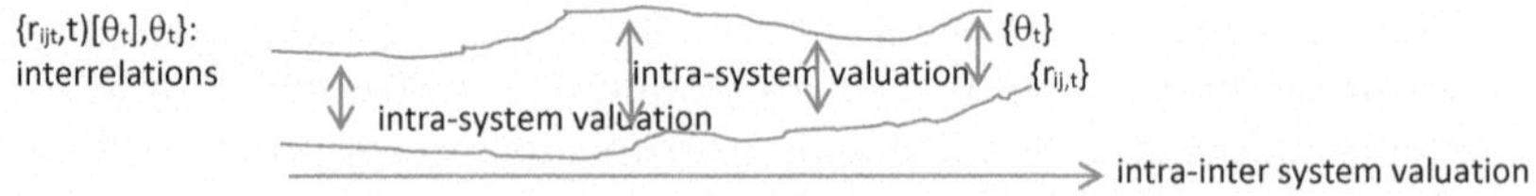

Fig. 7.2 Event-point evaluation across historical knowledge, space, and time evolution

established by such an equivalence between Tawhid as law and maqasid al-shari'ah as the right way in the light of Tawhid.

8. The subtle result of the equivalence between Tawhid as law and maqasid al-shari'ah as the right way of determining choices by discursive investigation is a sure way of replacing the debt instruments by the truly Islamic ones of matching money, financing, and the real economy in rounds of circular causation. The latter estimation and simulation method arises from the evaluation of the wellbeing function (maslaha), and its reconstruction into recommended forms. The maslaha index is thus appropriately determined for conceptual and applied use out of the sheer principle of organic unity of knowledge that arises from the ontological foundation of Tawhidi worldview.
9. The transdisciplinary interactive, integrative, and evolutionary learning model of Tawhidi methodology in continuums is the surest way of establishing the legitimacy of the Islamic financing complementarities within them and across the participative life-fulfilment development regime. Such is the regime that the Islamic complementary instruments are made to attain for the purpose and objective of wellbeing (maslaha) and its various elements.
10. The reading of the shari'ah reports points out a deepening focus of a shareholder model to access and raise rents in the globalization mercantilist agenda. Shari'ah as taken in differentiation with the Tawhidi methodological worldview has failed to configure a true Islamic transformation at any time and any place in history. An example is that of the Ottoman World in regards to shari'ah and its governance by central authority and with the objective of ruling over the Ottoman World for a full 600 years. The matter of integrating shari'ah with secular law of Europe during the reign of the Ottomans over 600 years produced an amalgam between these two branches law. The result was a secularization of shari'ah in a legal system called Kanun. Shari'ah then was found not to be developed enough to address the financial, taxation, and administration matters of the expanding Ottoman empire. Thus, there came in the integration of secular law into shari'ah by sometime replacing the latter. Besides, the kanun with certain power of secular law over shari'ah was found to be necessary to curb the oppressive powers of the Islamic clergy. It offered a way to overcome the problems posed by the extent to which Sharia depended on the interpretation of sources by the ulema, which had made legal standardization problematic.

As an example in the Ottoman World, the reign of Suleiman the Magnificent was a perfect example of this. Religion and thus shari'ah for the Islamic counterpart of kanun was brought under state control by the setting up of a hierarchical structure centred in the capital, Istanbul. All religious matters and thus the enactment of shari'ah within kanun had to be approved by the central government in Istanbul (Eldem, E. 4 (October 2013). "Balancing sharia: the Otoman Kanun", BBC Religion and Ethics, online).

The above narrative is a historical evidence of long standing to prove the role that shari'ah played in its secularization during the expansionist period of the Ottoman World, which was similar to a form of globalization of present times. The exemplar derived from the perusal of presently existing shari'ah reports proves a legacy of globalization focus as well. Just like the incapability of shari'ah at the time of the Ottoman World to address the problems of finance, taxation, and administration of a globalizing empire, so also today, the competition fervour to penetrate a mercantilist globalization agenda of the so-called 'Islamic' finance has caused an emersion of this programme to western modes of financing. The result has been negligence and incapacity of shari'ah to instil scholarship into the mode of developing a socio-scientific system of Islamic finance and development. Instead, the fervour of globalization has led into concerns with shareholder wealth formation and market expansion. In all of these global financial penetration agenda, the state etatism prevails across the Muslim world, just as the legitimation of the shari'ah in kanun had to regulated and governed by the central authority of the Ottomans in Istanbul.

11. Because of failure of shari'ah and its proponents in 'Islamic' finance, the debt instruments have predominated this field. This has caused a barrier to the development of a monetary-finance-real economy development financing model that the Tawhidi methodological worldview offers in the field of money and real economy with the replacement of debt instruments by direct trading financial valuation instruments of permissible development financial portfolio under the maqasid al-shari'ah choices in the light of Tawhidi methodological worldview of unity of knowledge and its induced world-system.

Multisystem and Multivariate Evaluation of Wellbeing Function (Maslaha) and Asset Valuation

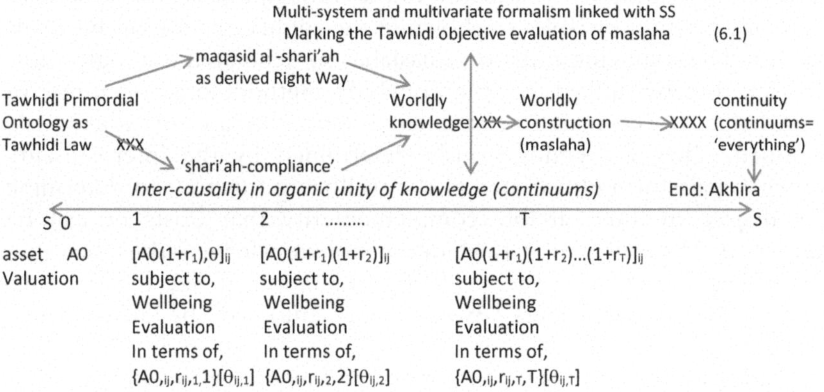

7.1

The inter-variable wellbeing evaluation across knowledge, space, and time dimensions is shown by the diagrammatic expression (7.1).

The introduction of multisystem and multivariate formalism out of Tawhidi methodological worldview of sustained continuity of unity of knowledge across all systems and variables in respect of knowledge, space, and time dimensions establishes the universality and uniqueness of such inter-causal relations. In symbolic terms, let $\{f_{ij}(\theta, x_{ij}(\theta))\}$ denote the inter-causal relations for *i*-systems, *j*-variables in multivariable and multisystems. θ = average$\{p$ lim $\{\theta_{ij}\}$ also denotes the linear simplification of the non-linear and complex expression of $\{\theta_{ij}\}$ - knowledge parameters across systems and multivariates. The $\{\{\theta_{ij}, \theta\}$ - knowledge parameters therefore represent the quantitative expressions for the conceptual forms of the wellbeing functions (maslaha), $\{W_{ij}(\theta_{ij}), W(\theta)\}$.

$W(\theta) = \prod_i W_{ij}(\theta_{ij})$, evaluated subject to inter-variables endogenous relations of $\{f_{ij}(\theta, x_{ij}(\theta))\}$, across i-systems, j-variables. The equivalent expression amenable to quantitative analysis is, $\theta = {}^{1/n}\sqrt{\Pi_i \theta_I}$; with each i-systemic $\theta_i = F_i(x_{ij}(\theta_i))$. The bold symbol denotes vector of all j-variables. Such quantitative evaluation of the wellbeing function, subject to circular causation variables, is shown in the applied section of this manuscript. The wellbeing function (maslaha) so generated is a logical formalism directly linked to the Tawhidi methodology of unity of knowledge. The estimated and simulated coefficients are not exogenously assigned. They are endogenously determined by the inter-causality between pairs of multivariates of the wellbeing function to determine the degree of inter-variable complementarity that exists or can be improved. The principle of Tawhidi unity of knowledge is thus the cause and effect circularity in the wellbeing evaluation, unlike the shari'ah-type maslaha coefficients exogenously enforced by institutional preferences.

These expressions yield the intertemporal functions noted in Fig. 7.2 as in the case of $\{r_{ij,\,t}, \theta_t\}[\theta_t]$ for the case of asset valuation intra-system and inter-variables denoted by $\{r_{ij,\,t}, \theta_t, t\}[\theta_t]$. The Tawhidi methodological way of estimating the rate of return uses the intertemporal vector, $\{r_{ij,\,t}, \theta_t, t\}[\theta_t]$ in asset valuation by the intergenerational terminal valuation model as shown in Fig. 7.2. This calculation is done across intra-system and inter-system in knowledge, space, and time dimensions. The shari'ah approach to riba or rate of return calculation is done in each intra-space as shown in Fig. 7.2. The intertemporal case of riba/rate of return is done along each of the intra-systems across time.

It can be proved that 'r' can be estimated by its own equation of the circular causation system in terms of the other variables, for example $(A0, \theta_t)[\theta_t]$, and further extended to multisystems if applicable. This is the intra-systemic case. For the inter-systemic evaluation of 'r', the estimation can be done by the equation, $r(\theta) = f_1(A0, \theta, t)$, and recursively with, $\theta = f_2(A0, r, t)$ as the quantitative form of the wellbeing function for the case of intergenerational valuation of assets. This method of evaluation of $\{r(\theta), \theta, t\}$ by circular causation intergenerationally is implied by Fig. 7.2. The measure of 'r' yields the rate of return or the interest rate as the time

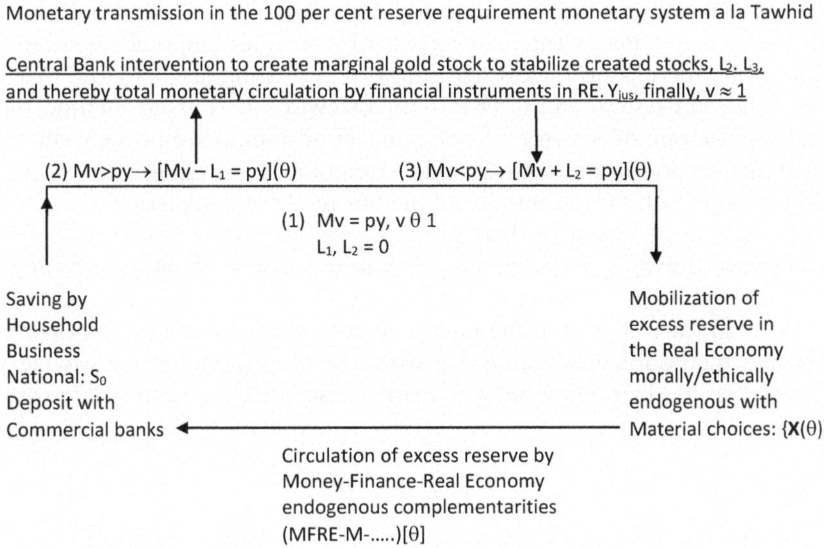

Fig. 7.3 Monetary transmission in 100 per cent reserve requirement monetary system of (MFRE)[θ] endogenous complementary relations

value of money in the valuation equations.[3] The valuation model derived from Tawhidi methodology of IIE existing in continuity of knowledge (thereby 'de-knowledge' as falsehood) in time. In every case of valuation, 'r' is determined 'nearest' to the point of occurrence of the event given by $\{A0, r(\theta), \theta, t\}$ across time along the sustainable continuous path of recurrence of the IIE learning processes SS.

The evaluation of the wellbeing function by the Tawhidi methodology establishes the difference in the measurement of riba by the intertemporal trajectory by evolutionary learning across intra-system and inter-system

[3]The intergenerational knowledge-induced model of asset valuation takes the form of terminal value TV, $TV(\theta) = A0(\theta) \times (1 + r(\theta))T$, with the corresponding circular causation relations between the variables of the vector, $\{A0(\theta), r(\theta), \theta, t\}$ for the intra-systemic case. Note that the entire period of valuation is linked up with TV by the geometric mean (or any other average) $r = \sqrt[1/T]{\prod_{t=1}^{T} r_t(\theta)}$, and the quantitative form of intra-system maslaha indexes given by, $\theta = F(A0(\theta_t), r(\theta_t), t)$. In the case of inter-system linkages 'r' is evaluated by the inter-causality between the variables of the vector $\{A0(\theta), r(\theta), t\}$. These cases of evaluation are shown in Fig. 7.2.

inter-causal linkages and the meaning given to riba by lateral addition across intra-systems, which is incorrect idea of intertemporal valuation. Contrarily, there is no logical derivation of evaluating assets by the riba-free rule intra-system and inter-system. Likewise, there is no method of deriving the rate of return under the impact of imminence of the wellbeing function and in the context of intertemporal knowledge framework. This is the reason for the absence of the 'nearest' event approach to evaluating '$r(\theta)$' in any form of shari'ah methodology, as is the otherwise the case with Tawhidi methodology of unity of knowledge governing 'everything'.

With all the analytics of the intergenerational asset valuation model of Tawhidi methodological genre, we write the corresponding knowledge-induced asset valuation model with market-institution decision-making as,

In reference to Tawhidi methodology, $TV\left(r(\theta),\theta\right)=\sum_{t=0}^{T}\left\{A_{ij,t}\left(\theta_{ij,t}\right)\times\prod_{t=0}^{T}\left(1+r_{ij,t}\left(\theta_{ij,t}\right)\right)^{t}\right\}$, with $A_{ij,\ t}$ denoting net asset value at every point of time, t = 0, 1, 2, ..., $T\{r_{ij,\ t}\}$ denotes (i,j,t)-specific rates of return, with $r=\sqrt[1/T]{\prod_{ijt} r_{ij,t}}$. Likewise, it is the case with $\{\theta_{ij,\ t}\}$, and the geometric average θ derived therefrom. TV denotes the asset valuation at the 'nearest' point of occurrence of the event, $\{A_{ij,\ t}(\theta_{ij,\ t}), r_{ij,\ t}(\theta_{ij,\ t}), \theta_{ij,\ t}, t(\theta_{ij,\ t})\}$, subject to the evaluation of the quantitative form of the wellbeing function, subject to the inter-variable circular causation relations. Among these relations are the estimated (positive) and simulated (normative) values of the 'nearest' value of rate of return 'r'.

In the absence of Tawhidi methodology of dynamic organic relations by unity of knowledge, maximization of the valuation model leading to determining the rate of interest (discount) is required. Thereby, the only maximum value of TV would be infinity, and thus indeterminacy in the case of time value of money as the discount rate or rate of interest. For allowing for other methods of asset valuation, as by the present value model and its various derivatives, shari'ah approach has remained silent on the choice of the appropriate method. Therefore, the knowledge-induced intertemporal concept of rate of return/interest rate remains absent. The riba rule fails to have a precise conception and quantitative determinacy. The shari'ah's openness to riba rule of asset valuation thus accepts the present valuation rule of determining the rate of return/riba as the internal rate of return. We can then write by using the Newton-Rhapson method of determining the positive rate of return as a root of the polynomial formed by the terms of the present value $PV(r)$,

$PV(r)=0 \Rightarrow \sum_{t=0}^{\infty} A_t \times X^t$, where $X = 1/(1 + i)$. 'i' denotes rate of interest (time value of money). A_t denotes time, t-periodic expected cash flows.

Newton-Rhapson method of solving the polynomial with roots '$X = 1/(1 + i)$' yields:

$X_{n+1} = X_n - (PV(X_n)/PV'(X_n))$, where X_n is a given value in the neighbourhood of solving for roots of the polynomial, $PV(X_n)$. $PV'(X_n)$ is the first derivative of $f(X_n)$. X_{n+1} is the Taylor Series iterated value of X_n.

The 'i' root calculated by the PV, and thereby internal rate of return method is clearly based on the riba methodology of time value of money. Besides, it is a rate calculated on deferred payments over time, which is a contrary postulate to 'nearest' point of possession (qabd) of any particular attribute of an asset that is conveyed by the existing value of the event, $\{A_{ij,t}(\theta_{ij,t}), r_{ij,t}(\theta_{ij,t}), \theta_{ij,t}, t(\theta_{ij,t})\}$, and thereby having no reference to a participatory reference to knowledge-induced market institution decision-making. In all such conditions of asset valuation by present valuation formula, as a consequence of the silence of shari'ah on estimating the rate of return/rate of interest on perpetuity, the present valuation and internal rate of return formulas are legitimated. The result is that the expected cash flows, which have in them the debt leverage over time, remain devoid of valuation at the 'nearest' point of possession of value (qabd). The measure of value in the 'nearest' point case is given by the change in the wellbeing function in terms of the vector value of change in the variables as elements, with respect to knowledge parameters 'θ'.[4]

[4] In the case of the wellbeing function, say $W(\theta) = W(x_1, x_2, \ldots, x_n)[\theta]$.

Here, $dW(\mathbf{x}(\theta))/d\theta = \sum_{i=1}^{n}[\{\partial(W(\mathbf{x}(\theta))/\partial x_i(\theta)\} \times (dx_i/d\theta)]$ measures the value of the choice, $\mathbf{x}(\theta) = \{x_1, x_2, \ldots, x_n\}[\theta]$, $i = 1, 2, \ldots, n$, in terms of the wellbeing function.

For a false variable $x_f \sim (\theta)$ in respect of the knowledge parameter 'θ', the negative effect (mafasid) is shown by $[\partial\{W(\mathbf{x}(\theta))/\partial x_i(\theta)\}]\ 0$. This can be referred to pure value of wellbeing (maslaha), pure 'de-value' of harm (mafasid), or a mix of maslaha and mafasid as codified by specific elements of $\mathbf{x}(\theta)$.

It is noted that in value calculation, the specific equation for quantitative value of wellbeing denoted by 'θ', this is a simplified linear approximation of the non-linear and complex form of $W(\theta)$ in terms of 'θ'. In the case of 'θ' as geometric average of the $\{x_i(\theta_i)\}$ – specific θ_i – parameters, we write, $W(\theta) = {}^{1/N}\sqrt{\prod_{j=1}^{N} \theta_{ij}}$. 'i' denotes column values of different $x_i(\theta)$ – variables. 'j' denotes N-number of observations of each ith-variable. In such non-linear and complex calculations either time series or survey methods of data collation can be used. These kinds of knowledge-induced and wellbeing-specific 'nearest' point valuation method are not any ingenuity of the shari'ah conceptions and instilling of analytical mind. Contrarily, these attributes are deeply ingrained in the vista of Tawhidi methodological worldview.

The absence of 'nearest' point evaluation of the knowledge-induced type in reference to the wellbeing criterion and the analytics of interrelational institution and market discourse results in inadequacy of the shari'ah-oriented approaches to debt instruments. Thereby, the grand generalized system problem of money, spending (finance), and real economy organic relationship as of the Tawhidi methodological application (see below) remains absent and, according to shari'ah, in all its different forms. It is pressing time now to recapitulate on the failure of orthodox economic scholarship and its overshadowed financial institutions, public and private leverage, and the so-called governed policy-making to forebode and reconstruct the future of global wellbeing. The continued unsustainability of shari'ah methodology and its methods, which are immersed in orthodox economic, financial, and socio-scientific thought that brought about the great financial crisis of the 1990s, is insecure future moorings for the global prospect under a distinctive mind–matter revolution under Islamic worldview.

Rob Johnson, president of the Institute of New Economic Thought (INET), wrote on the gathering storm of future uncertainties in a debt-burdened socio-economic system (Johnson, "10 Years After: A Retrospective on the Financial Crisis", info@ineteconomics.org, Oct 1, 2018) to erode human wellbeing: "Yet, these fallacies persist, from the academy to finance ministries to central banks. And as long as these wrong-headed characterizations of finance keep hanging over the design of the financial sector that is imposed on society, the pressures of excessive private debt and overly optimistic predictions of control will once again threaten to damage the world economy."

On the other hand, how does the Tawhidi methodological worldview offer solution to the debt crisis and to the complete proliferation of debt instruments of the shari'ah genre? The 'nearest' evaluation method of resource mobilization via the money-finance-real economy generalized system model is a pointwise (intra-system) and intertemporal (inter-system) knowledge flow activity. In it the objective perpetuates the socio-scientific goal of evaluating wellbeing (maslaha), subject to the inter-variable circular causation relations formed by the Tawhidi methodology of organic unity of knowledge. All this suggests that, resource mobilization of bank savings is continuously evaluated by the socio-scientific wellbeing objective criterion in the knowledge, space, and time dimensions. In the money-finance-real economy system of continuous flows of resources in financial and material forms, only a marginal residual saving remains unmobilized at the 'nearest' point of financial valuation. Thereby, the impact of exchange as trade within the resulting 100 per cent reserve requirements monetary system interrelating the central bank, commercial banks, and the market economy, resolves the problem of riba hangover and thereby debt instrumentation of financial flows. Likewise, the problem of risk and production diversifications results in reduction of unit risk via increase in stakeholding. The implication of life-fulfilment regime of socio-economic development inheres in all such ethico-economic reconstructions.

Failure of Shari'ah-Dictated Debt Instruments and Monetary Reform and the Tawhidi Alternative

The continued failure of shari'ah research to offer anything truly Islamic in the context of the monotheistic unity of knowledge as the ontology of the Tawhidi methodological worldview has held back the promise of an Islamic capital market in the ummah and global scene. Today in spite of the clamour around increasing market share by Islamic financial instruments and no intellectual contribution for an abiding methodological worldview, the statistical figures remain feeble. Then too the effort in expanding the market shares owes to government and rich shareholding strategies. Social participation in the development of the Islamic worldview is weak and eroding. In a recent media report completed in Turkey[5] it is researched to find the following result: "This study points out to a

[5] Kizilkaya, N. & Meric, Y. (2018). "Media Coverage of Islamic Economics Report", IKAM, Istanbul University, Istanbul, Turkey.

change, considering that Islamic economics was designed as an alternative paradigm and is regarded as an essential part of Islam's comprehensive worldview in the works of pioneers of this field. It is thought-provoking that Islamic economics, which is trapped mostly in the area of Islamic finance, is not seen as an alternative system by the media institutions, either."

To reinstate the importance of monetary, financial, and real economy circular causation linkages in the light of Tawhidi organic unity of knowledge, we now present the imminent model of monetary reform. This section is a continuation of the recent papers by Choudhury (2015, 2017) and Choudhury and Hoque (2017) on the topic of endogenous nature of money (Choudhury, 1997) and monetary transmission as micro-money in Islamic financial economics. These papers have projected an Islamic economic theory of endogenous money that pursue the financing of projects in the real economy by bringing such project-specific spending to attain several much wanted goals of economic stability and performance. Among these is the 100 per cent circulation of bank savings that banks otherwise hold back into the liquid form of spending.

A 100 per cent Reserve Requirement Monetary System (RRMS) is maintained in Islamic monetary transmission and circulation with appropriate methodology and policies. Consequently, debt reduces to zero; the market transformation by the diversification of projects, the portfolio diversification of financing instruments to mobilize monetary units as micro-money into projects, and the increase in stakeholding altogether make the real output level statistically elastic. Thereby, prices stabilize in the absence of all forms of interest rates. The end result is a generalized inter-causality among the endogenous inter-variable relations.

The 100 per cent RRMS is a logical consequence of the MFRE model of resource mobilization of all kinds. It was the nature of monetary system that was vouched by Imam Ghazali. Ghazali (trans. Karim, n.d.) wrote that any coin that was not supported by a certain amount of gold numeraire was not a valid coin in circulation. It therefore did not qualify as a resource-mobilizing instrument. Ghazali also believed that fluctuation of market prices of goods and services was caused as an act of God, and not by market demand and supply, as we theorize in economic study. Such a thought invoked knowledge derived from the Tawhidi ontological origin to induce market processes. To the extent that, market price fluctuation governed over exchange as trade, and thereby resource mobilization, the velocity of money circulation according to the quantity theory of money also fluctuates around a value of unity. A resulting automatic stabilization of the quantity of money to the market value of spending in goods and services causes a self-governing evolutionary equilibrium in all interrelated markets.

Consequently, Quantity of Money ⇔ Value of Spending (Finance) ⇔ Automatic Stabilization ⇔ Debt Instrumentation ≈0 ⇔ riba ≈ 0⇔ risk and production diversifications ⇔ self-governing socio-economic order ⇔ attainment of simulated balance in the wellbeing function. The symbol ⇔ means inter-causality between the various attributes shown here.

Endogenous Monetary Transmission in Islamic Financial Economics

The comprehensive picture of organic interrelations between money, finance, and real economy in the 100 per cent reserve requirement monetary system is shown by the following sequence of effects: $S(r/i,y) = M(r/i,y) \times v = p \cdot y \Rightarrow$ 100 per cent conversion of bank savings into money that is circulated fully via projects by financing instruments into the real economy. The symbols are defined as follows: $S(.)$ denotes bank savings as deposits that are held by banks as liquidity that is not mobilized into the real economy. '*r*' denotes rate of return in the real sector mobilization of savings and money circulating via financing instruments. '*i*' denotes rate of interest. '*r*' and '*i*' may be in nominal or real terms. '*y*' denotes real output. M denotes real quantity of money as a function of the real values of y and r/i (Cagan, 1989). '*p*' denotes the price level. '*v*' denotes the velocity of circulation of money. As the circulation of a quantity of money in the above-mentioned relations equates with bank savings liquidated into real economy financing (spending), then v tends to 1 as the signal of full mobilization of bank savings into productive spending.

Micro-Money Formalism in Islamic 100 Per Cent Reserve Requirement Monetary System

There are important social implications as endogenous ones linked with the economic ones of the above-mentioned relations. The fact of price stabilization with the perfectly elastic growth of the economy as a permanent phenomenon of the money, financing (spending), and real economy circular relations in respect of their endogenous inter-causal relations means that the economic integration of this type has feedback with a life-fulfilment needs economy and an increasingly privatizing economy. In the case of the macroeconomic general equilibrium system, this kind of adjustment is equivalent to an elastic form of non-inflationary economic growth and the complementary relations between aggregate demand with predominating spending (fiscal) stimulation and monetary policy induction (Friedman, 1953).

The project-specific expansion of monetary transmission complementing with the financing activity causes real output expansion. The results then are the market dominated transformation with public sector discursive coordination. Thereby, the real output expands along the intersection of the resulting aggregate demand and aggregate supply curves. Such a movement causes an elastic increase in productive employment and price stability. In the Keynesian jargon, the full impact of the income multiplier is realized along such non-inflationary growth paths. Finally, price stability as a result also causes simultaneously a stability in (r/i) price relative.

Monetary Transmission Model of Islamic 100 Per Cent Reserve Requirement Monetary System

In Islamic economic and real economy theory, the complementary coordination of money, finance, with the above-mentioned inner results in the variables causes reduction in 'i' even as the circulation of money, that is monetary transmission between banks and the real economy, enhances productive transformation. Figure 7.1 points out the nature of money and monetary transmission between the banks and the real economy in respect of liquidating the bank savings into productive mobilization by spending. The well-known theory of Money (M)-Commodity (C)-Money (M) model (Heilbroner, 1986) is now extended to $\{(\text{MFRE})_{\text{transmission 1}} - (\text{MFRE})_{\text{transmission 2}}\ \text{etc}\}(\theta)$.

Symbolizing the Basic Model of Tawhidi Unity of Knowledge and its Conceptualization in Money, Finance, and Real Economy Endogenous Interrelationship

Symbol 'θ' denotes consciousness parameter of knowledge based on the episteme of unity of knowledge. In Islamic epistemological worldview, the derivation of 'θ' is discursively determined by way of the foundational reference to the Islamic primal ontology of oneness of God (Tawhid). This premise is explained by the precept of unity of knowledge in endogenously circular causal interrelations between the good choices identified by their multidimensional variables. Briefly, we write this foundational knowledge-deriving epistemological methodology from the primal ontological roots as follows:

$\Omega \rightarrow_S \rightarrow \theta = p \lim \{\theta$ by discourse interactively with the learned ones$\}$ (Qur'an, 4:59).

Ω denotes the primal ontology as the Qur'an, emphasizing oneness of Allah represented as unity of knowledge by organic interrelations between the permissible choices while avoiding the impermissible ones (Qur'an, 36:36).

'*S*' denotes the transmission of the monotheistic knowledge in the Qur'an into worldly specifics denoted by 'θ'. 'S', called Sunnah, comprises the teachings of Prophet Muhammad, who Allah appointed by his Qur'anic command as his chosen messenger for enlightening access to 'everything'. The meaning of 'everything' could be noted in Barrow (1991).

$\theta = p \lim \{\theta\}$ is the consensual derivation of unity of knowledge. It can be ordinally represented in the derived model of (MFRE)[θ] by discourse in reference to the law of monotheism explained by the episteme of unity of knowledge as an organic process of knowing in the Qur'an and Sunnah.

The knowledge of evolutionary discursive system of multidimensional evolutionary learning processes of diverse world-systems, all induced by knowledge arising from the episteme of unity of knowledge and contrarily from the differentiated worldview of disjoint diversities not guided by $\Omega \rightarrow_S \rightarrow \theta$, both carry on inducing the generality and details of the specific world-systems under study. In this chapter, the specific world-system under study under the ontological and epistemological determination by $\Omega \rightarrow_S \rightarrow \theta$ is the (MFRE-...)[θ] complementary problem under the guidance of 100 per cent reserve requirement monetary system.

The evaluation of the effectiveness of the organization and realization of the (MFRE-...)[θ], as explained above, is carried out by the wellbeing function, referred to in the methodology of $\Omega \rightarrow_S \rightarrow \theta = p \lim \{\theta\}$ as maslaha. The evaluation is made by estimation and simulation of this function, subject to the system of endogenously related inter-variable equations of circular causation relations in the light of the episteme of unity of knowledge, and vice versa, as established in $\Omega \rightarrow_S \rightarrow \theta = p \lim \{\theta\}$. The maslaha function as wellbeing objective criterion is denoted by W(MFRE,θ)[θ] with the inner variables.

Monetary Transmission in the 100 Per Cent Reserve Requirement Monetary System A La Tawhid

A brief explanation of monetary transmission circulating in the 100 per cent reserve requirement monetary system is as follows. The initiation of savings is by productive activity in the national income, by way of which savings S_0 are deposited in the commercial banks. In the 100 per cent reserve requirement system, the savings become deposits in banks ready to be circulated as money via financial instruments (F) into the real economy, RE. Now there are three different scenarios in such a circulation of money by financial instruments into the real economy. Firstly, the full amount of savings as deposits can circulate in the real economy by financing instruments aimed at specific projects. This kind of monetary circulation results in micro-money pursuing projects. Thereby, with the full mobilization of savings as deposits as money and financing is the feedback consequence of v tending to unity. The micro-money project-specific circulation causes the otherwise case of fractional reserve requirement to withhold all the required deposits for the full circulatory purpose of spending by micro-money. The underlying episteme of unity of knowledge is thus implied to prevail in sustaining this complementary or participatory organic relationship between money, finance (spending), and real economy [MFRE](θ).

This dynamic is represented by the parameterization of the unity of relations by (Choudhury, Pratiwi, Hossain, and Adenan in Toseef, forthcoming). The origin of such parameter-based inter-causal relations between the activities along with their inner variables are denoted by (MFRE)[θ]; and M = *py*, with interrelated $M = A\times\prod_{i=1}^{n} M_i^{ai}(\theta)$; $\left[p{\cdot}y = \text{spending} = B\times\prod_{i-1}^{n}(p{\cdot}y)_i^{bi}\right][\theta]$. Both of these expressions apply in terms of project-specific micro-money and real-economy spending in them according to Islamic recommended choices (1 to *n*). Thus, the context of θ - induction appears.

The other cases of monetary transmission shown in Figure 7.3 are, firstly, of excess quantity of money adjusted by a withdrawal of L_1 of quantity of money to restore the equation of exchange: ($Mv > py \rightarrow [Mv - L_1 = py](\theta)$). The other case is of excess demand for money with adjustment thereby with injection of loanable fund by the central bank of L_1. In every case, the adjusted equation of quantity theory of money and spending is restored. The result then is identical to the 100 per cent mobilization of money into the real economy via financial instruments and banking institutions. The central bank applies a marginal quantity of gold as numeraire to stabilize the currency value of all monies in circulation, including L_1, L_2, and the rest.[6]

The complementary dynamics of [MFRE](θ) prevails always. The result in monetary transmission then is 100 per cent reserve with commercial bank to match monetary flow through financial instruments with real economy activity. This is the meaning of 100 per cent reserve requirement monetary system. The central bank thereby ceases to have monetary governance on monetary transmission. Only marginally there exist the management of L_1 and L_2, and the overseeing and guidance of the states and policies of the economy with micro-money. In such money-finance-real economy interrelations, the central bank establishes a discursive environment of consultation and policy guidance with commercial banks (Islamic banks), financial institutions, the real economy agencies, and all participants in economic activity.

A Focused Review of the Literature

In the seminal works of Maritain, 100 per cent reserve requirement monetary transmission of the [MFRE](θ) type appears. Maritain's contribution on a nature of money recommends unlimited tokens representing money for circulation in the real economy, so that all individuals and households would benefit from these tokens by exchange of goods and services. Monetary control by the central bank, commercial banks, and other monetary authorities would disappear. In 100 per cent reserve requirement monetary system with the gold standard as a stable numeraire, the circulation of micro-money is similarly recommended to be unlimited so as to be circularly related with real exchange in goods and services.

[6] The dynamics of monetary transmission under 100 per cent reserve requirement monetary system imply that there is an inverse relationship between creation of gold (G) and trade (py). Thereby, gold creation comes to end as the adjustments in limiting amounts of L_1 and L_2 restore the 100 per cent mobilization in (MFRE)[θ]. We write:

$L_1 - L_2 = \Delta M = \Delta py = G$. We rewrite this expression as, $\Delta py/py = G/M$. That is, currency value of money in circulation, py, equals G/M. Now as monetary transmission increases, M increases in quantity, G declines in volume. Thereby, $G/M = a$, as a declining parameter. Change in currency value, $g(py)$, is therefore stabilized around 'a' the total amount of money in circulation. That is $G = a \times M$, with G declining asymptotically, as 'a' and M movement establish a rectangular hyperbola.

Thereby, the wellbeing effect will be optimized and asset-backing by gold would be minimal in volume. Gold would be required by the central bank only to stabilize the circulatory currency value and the residual amount of excess reserves in the central bank (Fig. 7.3).

The praise for expansionary monetary supply for meeting the demand for spending in goods and services in the real economy was also proposed by the Austrian School of Economists (Yeager, 1997; Hayek, 1999). The unit of micro-money was named as Unit. It was thus the numeraire to match up with a unit value of spending in goods and services in the real economy. On the side of the wellbeing contribution of such micro-money Schumpeter's (1961) entrepreneurial growth model is an example of micro-money being in high supply to finance sustainable microentrepreneurial projects.

In recent times, it is surmised by hind sight that the monetarism of an easy rather than a counter-cyclical contraction in the quantity of money into the real economy could have stemmed the tide of the Great Depression. In a growing economy, the complementarities between fiscal and monetary policies can be best fitted for non-inflationary economic growth.

In all such states of [MFRE](θ) model within the regime of 100 per cent reserve requirement monetary system, there remains the consistency of the effect of declining interest rates on the complementarities between money, finance, and real economy. Furthermore, the expanded release of monetary supply to endogenously support entrepreneurship, sustainability of non-inflationary economic growth, and thereby wellbeing. The early work of Friedman (1960) pointed out the possibility of such a holistic possibility of monetary arrangement.

The [MFRE](θ) model along with its early prototypes in the history of money, monetary policy, and institutional control over quantity of money for the real economy suggest that this model is consistent with the state of a productive and ethically sustainable economic regime with wellbeing as monetary and financial resources increased to complement with the real economy. Thereby a circular causation exists between the dynamics of money, finance, and real economy interrelations and the phasing out of all forms of interest rates causing release of resources for sustaining the holistic consequences of the wellbeing criterion. This yields the situation whereby a dynamic evolutionary learning by endogenous interrelations between the inner variable of [MFRE](θ) type monetary transmission model is attained (Hayek, 1990).

Economic Stability by Means of the Relationship Between [MFRE](θ) and the Wellbeing Function

Economic stability caused by the [MFRE](θ) is explained in the case of the theory of monotheistic unity of knowledge in Islamic economics and finance by evaluation of the wellbeing function called the maslaha. Maslaha as wellbeing function parameterizes and explains the degree of pervasive positive complementarities that ought to exist between the variables denoting the good things of life and avoiding the negative complementarities with the respective variables signifying the undesired things. The maslaha objective function also has the properties of inter-variable complementarities of the two kinds. It is continuously differentiable because of the same nature of the learning parameter derived from the ontological and epistemological foundations of unity of knowledge of the monotheistic methodology. That is, $\theta = p \lim \{\theta\} \leftarrow (\Omega, S)$. Due to the properties of interactive, integrative, and evolutionary learning of θ in the maslaha function (Choudhury, 2013), the maslaha function has non-optimal and non-steady state properties. Only evolutionary equilibriums caused by the force of 'θ' in generating estimated and simulated complementarities apply.

The ontological-epistemological (Spencer, 2000) origin of 'θ'-induced maslaha function also gives rise to endogenous circular causation relations between the maslaha variables. Thereby, the maslaha function is evaluated by estimation and simulation of coefficients in the 'as is' and 'as it ought to be' states of the problems under study in the light of pervasive complementarities between variables subject to the system of circular causation relations in evaluating the maslaha function.

Conclusion: Summarizing the Properties

One vector of selected variables in the [MFRE](θ) complementary problem in 100 per cent reserve requirement monetary system: $\mathbf{x}(\theta) = \{m, y, f, p, e, r/I, \theta\}[\theta]$, with induction of all variables by 'θ'. The variables are, 'm' as real money. 'y' as real output, 'f' as vector of financial instruments conducive of MFRE complementary relations in the circular flow of goods and finances, 'p' as output price signifying level stability, 'e' as employment, 'r/i' as relative between rate of return and interest rate, 'θ' as knowledge parameter inducing all these variables in terms of their inter-causal complementarities.

The conceptual maslaha function, subject to circular causation relations between the variables is $W(\theta, \mathbf{x}(\theta))$.

The circular causation relations between the variables are, $x_i(\theta) = A + \sum a_j(\theta) \times \ln xj(\theta) + a(\theta) \times \ln(\theta)$;
$i, j = 1, 2, \ldots, n$; $i \neq j$. Except for the last term, the θ - inductions of the variables and coefficients signify the sequential data of the variables corresponding to the θ - variables.

The quantitative form of the wellbeing function that can be evaluated (estimated in 'as is' state; and simulation in 'as it ought to be') is, $\theta = B + \sum a_i(\theta) \times \ln x_i(\theta)$. This equation provides the sequence of $\mathbf{x}(\theta)$ - variables that are estimated (simulated) as predictor variables from the circular causation estimated (simulated) results in the variables.

The full description of the evaluation system of the maslaha function is given by,

Evaluate

$$W(\theta,\mathbf{x}(\theta)) = \text{Const.} \prod x_i^{\alpha_i} \times \theta^{\beta}.$$

This is converted to natural logarithmic form, $\ln W(\theta,\mathbf{x}(\theta))$

$$= \text{Const.} \sum \alpha_i \times \ln x_i + \beta \times \ln \theta \qquad (7.2)$$

subject to the circular causation equations in natural logarithmic form,

$$\ln x_i(\theta) = A + \sum a_j(\theta) \times \ln x_j(\theta) + a(\theta) \times \ln(\theta). \qquad (7.3)$$

The quantitative form of the maslaha function to evaluate is,

$$\begin{aligned} &\theta = B + \sum a_i(\theta) \times \ln x_i(\theta) \\ &i, j = 1, 2, \ldots, n;\ i \neq j \end{aligned} \qquad (7.4)$$

Economic stability attained by the model of unity of knowledge generated by pervasive complementarities between the inner variables of [MFRE](θ) endogenous circular flow interrelations is explained further as follows. By some amount of simplification of expression (1) in mathematical differentiation form, we write in the form of rate of change, g(.) with respect to 'θ' in view of degrees of inter-variable complementarities:

$$g(W) = \sum_i \alpha_i \times g(x_i) + \beta \times g(\theta) > 0 \qquad (7.5)$$

Now because every variable is positively complementary with θ - value, therefore, economic stability in the midst of ethically induced variables of the [MFRE](θ) model is implied.

The explanation of economic stability is made by the ontology, epistemology, and phenomenology of Tawhid as law in respect of the Qur'anic meaning of monotheism and its methodological representation in terms of unity of knowledge between recommended complementary variables. Yet this result is not true of the shari'ah-based framework of Islamic economics and finance. Two examples are given here to establish the inability of shari'ah, fiqh, and fatwa's misguided claim of economic stability. This failure is reflected in the ISRA shari'ah-based report on Islamic commercial law.

Firstly, we take the case of the absence of ijara-type pricing, such as rental in ijara (assets acquired by deferred payments), premiums (as of takaful), deferred payments (as of mark-up pricing by murabaha), and risk and return pricing on credit sale (as in the case of twarruq, innah, and istisna). There are also the many cases that are daily conjured up by the so-called 'shari'ah-compliance' legitimation of Islamic financial instruments.

All such deferred rental kinds of rates as prices on financial instruments, and thereby their pricing relations with goods and services in market exchange, are indeterminate. This problem is true at the initial point of a mutual contract, unless such prices are forward-looking prices that can be realized only 'nearest' to the point of exchange transactions. Forward-looking pricing is tantamount to interest rate (riba), for it remains undetermined at the present time of a contract. Thereby, real economic transactions remain equally undetermined in valuation. On the other hand, if goodwill of donations is used to form financial capital, this is not feasible under continued conditions of uncertainty and financial limitations of groups and individuals as donors for free-ridership. Feasibility and sustainability demand organized forms of pricing for risk, return, and transactions in the intertemporal sense. No formula has been derived according to shari'ah to resolve this critical outstanding problem of forward-looking ijara-type pricing. There is no evidence of shari'ah instruments regarding their stability role in financing during global financial crises, as otherwise claimed. The only other way how this could happen is by relying upon large inflow of rich shareholders' funds. Such a case was implied in the greater than unit value of Finance/Deposit ratio in shareholding of Islamic banks during the financial crises.

References

Barrow, J. D. (1991). *Theories of Everything, the Quest for Ultimate Explanation.* Oxford: Oxford University Press.

Cagan, P. (1989). Monetarism. In J. Eatwell, M. Milgate, & P. Newman (Eds.), *Money, the New Palgrave* (pp. 195–205). New York, NY: W.W. Norton.

Choudhury, M. A. (1997). *Money in Islam.* London: Routledge.

Choudhury, M. A. (2013). *Handbook of Tawhidi Methodology: Especial Reference to Islamic Economics and Finance.* Jakarta, Indonesia: Trisakti University Press.

Choudhury, M. A. (2015). Monetary and Fiscal (Spending) Complementarities to Attain Socioeconomic Sustainability. *ACRN Journal of Finance and Risk Perspectives, Special Issue of Social and Sustainable Finance, 4*(3), 63–80.

Choudhury, M. A. (2017). The Future of Monetary Reform and the Real Economy: An Islamic Problem of Exchange Versus Interest. *ACRN Journal of Finance and Risk Perspectives, Special Issue of Social and Sustainable Finance, 6*(4), 37–52.

Choudhury, M. A., & Hoque, M. N. (2017). Comparative Islamic perspectives in Money, Monetary Policy, and Social Wellbeing. *Journal of Economic Cooperation and Development, 39*(1), 143–162.

Choudhury, M. A., Pratiwi, A., Hossain, M. S., & Adenan, F. (forthcoming). In T. Azid (ed.), *A Relational Wellbeing (Maslaha) Index Of Gender Development In Socioeconomic Development Sustainability*, Routledge, London.

Friedman, M. (1953). Monetary and Fiscal Framework for Economic Stability. In *Essays in Positive Economics* (pp. 133–156). Chicago, IL: University of Chicago Press.

Friedman, M. (1960). *A Program for Monetary Stability.* New York, NY: Fordham University Press.

Hayek, F. A. (1990, reprint). The Use of Knowledge in Society. In M. C. Spechler (ed.), *Perspectives in Economic Thought* (pp. 183–200), New York, NY: McGraw-Hill.

Hayek, F. A. (1999). Intertemporal Price Equilibrium and Movements in the Value of Money. In *Good Money, Part One: The New World* (pp. 186–227). Chicago, IL: University of Chicago Press.

Heilbroner, R. L. (1986). The Ideology of Capital. In *The Nature and Logic of Capitalism*, Chapter 5, New York, NY: W.W. Norton.

Schumpeter, J. A. (1961). *The Theory of Economic Development.* Cambridge, MA: Harvard University Press.

Spencer, N. (2000). On the Significance of Distinguishing Ontology and Epistemology. Retrieved February 2018, from https://ethicalpolitics.org/seminars/neville.htm.

Yeager, L. B. (1997). *The Fluttering Veil, Essays on Monetary Disequilibrium.* Indianapolis, IN: The Liberty Press.

CHAPTER 8

The Ontology of Tawhid as the Law Applied to Specific Issues

The Worldview Beyond a Materialistic Globalization Concept

In the light of the Tawhidi model of unity of knowledge, the overarching knowledge-induced model carries $\{\theta\}$ into the intersected set, $S(\theta)=\bigcap_{i=1} S_i$. By the topological nature of the functional relations between these sets, each component of the individual sets follows the same unitary relations caused by the $\{\theta\}$ - variables. Such member variables can be characterized as interactively integrated multimarkets, financial integration, technological innovation, and institutional and inter-country coordination that the $\{\theta\}$ - values induce as parameters signifying unity of knowledge.

The globalization process is now defined by the interaction and integration leading to evolutionary learning that takes place by the induction of $\{\theta\}$ - values on an epistemic scale. θ is such as it interactively integrates the diverse sets comprising economy, society, and science and technology across the diverse extension of organic unity caused by the epistemic embedding of the special kind of knowledge flow.

One would now consider from where such a special kind of unitary knowledge can arise in respect of globalization. In mainstream economic reasoning, there is permanence of the postulate of competition and self-interest that dominates; and because resources are assumed to be scarce, therefore, there must always be opportunity cost and marginalist trade-off between the S_i-alternatives. In the end thereby, the selected variables, their

M. A. Choudhury, *Tawhid and Shari'ah*,
https://doi.org/10.1007/978-3-030-49087-4_8

relations, and the representative entities must be independent of each other by methodological individualism as the common attribute (Underhill, 2003). The economic and political arenas interplay in this theatre of inherent conflict for power and control. This is a complex process of reality 'as is'. Yet it is a perpetual process of conflict that takes the S_i apart eventually to form dissociative separate wholes of the end state of methodological independence and individualism. Consequently, $\{\theta\}$ belongs to such an underlying epistemology and cannot therefore perform the normative expectation of interactive integration by the process of evolutionary knowledge premised in unity of knowledge. Under such an epistemology, $S(\theta) = \phi$, $\theta \in$ Rationality derived from Rationalism. This property contradicts monotheistic law of oneness and its organization of the organically unified world-system.

The above-mentioned properties of neoliberal and neoclassical social theory now prevails in Islamic economics. James Buchanan (1999, op cit., p. 391) writes on the exclusiveness of moral and ethical values in economics: "Concomitant with methodological individualism as a component of the hard core is the postulate of rational choice, a postulate that is shared over all research programs in economics."

At this point the discursive element of Tawhidi law in constructing a world-system in its generality and details enters the theme of globalization. In the light of the organic connectedness of the organic unity of knowledge by Tawhid, the theme of globalization invokes socio-scientific epistemic knowledge dissemination that through its intellection at every level enters the life-blood of the new world-system. Such a call was tried but did succeed because of the neoliberal and conflict episteme. Examples here are legion stretching from the idea of a New World Order (Choudhury, 1991) to that of One-World Currency (Mundell, 2000). The quest for the overarching coexistent worldview is therefore alive in both thought and action. The functional ontology of the unitary learning model is therefore a reality not Utopia.

Likewise, from the precept of the unified world-system of monotheistic law arises the determination of proper human resource development. There is also the choice of appropriate technology. Both of these practices are required to serve the purpose and objective of the basic needs regimes of development centred on poverty alleviation and social justice. It is along such unitary knowledge-centred worldview that the theme of globalization can acquire its appropriate human meaning. It is along this line that the extended idea of maqasid al-shari'ah in the light of Tawhid as the primal ontological law works in the multiverse domain of socio-scientific reality. This is the common arena where all nations can meet. The experience was exemplified by the Earth Summit in Rio. It is being pursued in the philosophy of multireligious dialogues that is sponsored by the Papal project in the Vatican.

Contrarily, the traditional idea of maqasid al-shari'ah fails to have such a meaning on the theme of globalization. Recent experience of the Muslim world and capitalist globalization has shown the Muslim world joining the neoliberal race to the bottom. No development along the lines of basic needs regimes could be defined. Human resource development is oriented to follow the neoliberal idea of human capital with all its ramification of neoclassicism. Technological choice is never innovative, always borrowed and imported from the industrial west. These predicaments cause immense financial and social dislocation in Muslim societies. These and many more attributes are the growing discontent of the Muslim world, where the voices of the appointed institution of shari'ah is allowed to serve self-seeking ends of economic power and politics. Regarding such Muslim predicaments in the face of a defunct usul al-fiqh (Islamic jurisprudence) Asad (1987) and Sardar (1984, 1988) have written.

In summary to this section, the extended idea of maqasid al-shari'ah within the primacy of Tawhid as law contrasts the neoliberal idea of justice and balance of relations and with the traditional idea of the shari'ah. The example of globalization on this matter can be extended to critical investigation of other issues within the socio-scientific domain. The next example of application of the generalized system meaning of Tawhid as law to maqasid al-shari'ah at-Tawhid is to the problem of debt and debt write-off.

Debt Instruments and Asset Valuation According to Shari'ah

Debt in asset financing and valuation is measured by the excess of investment demand over the available funds. Debt has its atemporal and intertemporal concepts. The atemporal concept of debt applies to the unpaid excess of investment demand loaned or contracted out to meet the excess demand for loanable funds. Intertemporal implication of debt is periodic payment to write off the unpaid amount of the loanable fund to meet the excess demand of investment.

An asset is valued in terms of its cash flows by a suitable method. In mainstream financial capitalization, the present value of the net cash flows based on intertemporal unpaid amount of debt outstanding is used. Another method is to evaluate the terminal value of the paid and unpaid debt outstanding at any given period of time. The usual rate of capitalization of debt in mainstream financial valuation methods is the rate of discount shadowing the rate of interest. This is the case for two reasons. The cash flows of the future are not known even in the probabilistic valuation of expected returns from investment. Consequently, the discount rate cannot be ascertained before the time of expected cash flows 'nearest' to the point of occurrence. Besides, in the terminal case, the valuation can be done 'nearest' to the occurrence of cash flows. The rate of valuation remains unknown ahead of time. In mainstream finance, any debt outstanding from loanable fund is subject to opportunity cost that is included in the estimation of the discount or future rate. Opportunity cost of such debt outstanding is equivalent to interest rate.

The only way for non-interest-based valuation of cash flows net of debt in Islamic economics is to measure the expected value of debt overhang, and thereby the net cash flows 'nearest' to the point of occurrence of the net cash flows. Such a form of debt liquidation involves redeeming capital gains out of an increasing amount of residual cash flows net of debt obligation. Consequently, the rate of return, although measured 'nearest' to the point of occurrence of cash flows, is expected to be on an upward trend. The net cash flows valuation of assets is thereby applicable only finitely over time.[1]

In this last case of net cash valuation, the intertemporal valuation is carried over time, with each time period enacting a fresh contract of sharing risk and return. In Islamic finance, the tradable nature of real assets against which risk and return are evaluated is done by the financial instruments of profit-sharing (mudarabah), equity-financing (musharakah), and cost-plus pricing (murabaha).

[1]Terminal value (TV) of net cash flows over time at rates of return r, (NCF(t);

$$TV\left(r,NCF(t) = (1+r)^t \cdot NCF(0)\text{done periodically,}\right. \tag{8.1}$$

$$TV\left(r,NCF(T) = \sum_{t=0}^{T}(1+r)^{T-t} \cdot NCF(T-t)\right) \tag{8.2}$$

done on futures market valuation. For this, r is estimated at the present time of investment or over time. In the first case, Islamic shari'ah approves net cash valuation but the issue of debt is debated. In the second case, the valuation by 'r' is equivalent to interest-based valuation.

By the generalized system model of Tawhid, the risk and return can be further diversified by secondary instruments. Now if joint venture funds are contracted at every point in time across diversity of financial instruments, then on a continuous scale and with many shareholders, stakeholders, and joint venture projects, debt is converted into payments by returns in equity. The result is a continuously transformation of debt into equity finance. Debt thereby ceases to exit by the presence of shares in joint ventures. Joint ventures replace debt across time. The valuation result then is to capitalize the future evolutionary returns on recontracted financing modes of joint venture financing. Such recontracted nodes of decision-making form the learning point. Changes by financing diversity in such nodes of decision-making form the evolutionary learning points of the type we have explained in the generalized system model of interaction, integration, and evolution based on the pairing concept of Tawhidi methodology.[2] It can now be shown (see footnote) that debt and debt-financing cannot exist in Islamic financing. Thereby, no notion of interest rates (riba) as a present price for the cost of deferred payments like debt can exist.

Debt is revoked from the Islamic participatory financing model by the Qur'anic precept of participation. The resulting Tawhidi methodological implications result in a logical formalism for such a model. However, in the traditional usul al-fiqh approach to interpretation of debt the asset

[2] According to the generalized system model of maqasid al-shari'ah, the evolutionary learning effect of joint ventures across diversification of projects (output) and participatory shareholders (stakeholders) takes place. Such is the conceptual principle of participatoriness in the Tawhidi law. It is specified to participatory financing instruments cancelling debt by joint ventures and continuous evolutionary learning signified by $\theta \in [(\Omega, S) \rightarrow \{\theta*\}]$.

$$\text{Now,} TV\left(r, NCF(t)[\theta] = \{(1+r)^t \cdot NCF(0)\}[\theta];\right. \tag{8.3}$$

$$TV\left(r, NCF(T)[\theta] = \sum_{t=0}^{T}\{(1+r)^{T-t} . NCF(T-t)\}[\theta]\right) \tag{8.4}$$

In each of these two cases, $NCF(t)[\theta]$ = (Return − Investment)$_t[\theta]$ = (Return − Joint Venture Capital)$_t[\theta]$. But $JVC = \rho_t(\theta). R(\theta)$. Therefore, $NCF(t)[\theta] = (1 - \rho_t(\theta)). R(\theta)$. Now if we include the rate of return 'r',

$$NCF(t)[\theta] = (1 + r_t(\theta)) \cdot (1 - \rho_t(\theta)).R(\theta) \approx (1 + (r_t(\theta) - \rho_t(\theta))) \cdot R(\theta). \tag{8.5}$$

Over time and $\{\theta\}$ - values, the rate of return equals the participatory financing rates of return. Then, $r_t(\theta) = \rho_t(\theta)$. Hence, $NCF(t)[\theta]$ = [R]eturn. Debt is then non-existent. It is, therefore, a misunderstanding to have debt and debt-based valuation in Islamic financing models under the evolutionary learning context of the generalized system model.

valuation allows for one-third of total loanable capital to be held in debt. There is no Qur'anic legitimacy for such a shari'ah rule (fatawa). In the end we deduce from the discussion here that debt cannot be avoided, and thereby shadow interest rate hangs on in the conventional shari'ah model of Islamic financing. The true Qur'anic and sunnah rule on the avoidance and paying off of debt at the earliest point in time is not abided by in this kind of an approach to Islamic financing.

The Tawhidi methodological worldview that can replace debt by a continuity of joint ventures by purely participatory financing instruments is not to be found in the shari'ah debt model. The context of the generalized system model of Tawhid applied to the valuation of financial assets in the absence of debt invokes the non-null and expanding core of the $S(\theta)$ expression by virtue of the evolutionary learning effect of $\{\theta\}$ - values. The sets on finance, institutions, society, and scientific diversity towards increasing productivity in projects with higher degrees of risk-diversification provide the intersecting and enhancing field of IIE-type participation.

Money–Real Economy Relationship According to the Generalized System Model of Maqasid Al-Shari'ah

Another example of the Tawhidi principle of complementary participation playing its role in the intrinsic nature of participative organic unification in the economy, finance, society, and science synergies is manifest in the money and real economy organic relations in the light of the evolutionary learning model of unity of knowledge. In all these interactive, integrative, and evolutionary categories, science is represented by the moral and ethical induction of information technology. Such emergent IIE process model of money and real economy is brought for formalism, study, and analysis within the non-linear, and complex system analysis, which the simple and isolated financial models cannot perform (Choudhury, 2013).

On a comparative examination, in neoliberal economic theory capitalism presents a relationship between money and the economy that remains indelible in all the relations, preferences, practices, and policies involving the economy, markets, and institutions. Heilbroner (1985) explained this inherent causality between money, capital formation, and the economic transformation in terms of Marx's MCM relationship. MCM can be explained briefly as follows:

A quantity of money (M) as fiat money is commoditized (C) through the process of saving leading to capital accumulation by interest rates. The savings then enter the time-lagged production process to cause the money held through savings to be mobilized partly into the real economy while the greater part of the savings continues in liquid form, as long as interest rates as the capitalist notion of return on money remain attractive to the savers. This process of commoditization of money followed by the emergence of more money can be explained by the simplified cycle,

$$
\begin{aligned}
&\text{M0(initial quantity of money)}\\
&= \text{M01(real money)} \qquad + \qquad \text{M02(paper money)}\\
&\qquad\downarrow \qquad\qquad\qquad\qquad\qquad \downarrow\\
&\rightarrow \text{savings and interest}\\
&= [\text{C01 (real economy)} \qquad + \text{C02(saving in financial outlets)}]\\
&\qquad\downarrow \qquad\qquad\qquad\qquad\qquad \downarrow\\
&\rightarrow \text{M11(monetary growth by real returns)} + \text{M12(monetary growth by interest rates)}.
\end{aligned} \tag{8.6}
$$

In the above schema, the part represented by **M01 → C01 → M11** is a purely resource mobilization channel that links money to the real economy. By this portion of the economic transformation is realized the participation of productive labour in economic activity. But this sequence also requires spending in the good things of life in order to qualify for both purposeful work efforts and mobilization of other productive factors that are complementary to one another. Through complementarities, participation increases in the moral and purposeful social and economic contexts. This causes the relevant form of productive technological transformation, empowerment, and growth of entitlement to take place. In this context, savings represent resources withheld (Ventelou, 2005). Continuous liquidation of savings into the real economic comprises resource mobilization pairing as complements, money, and real economy by the production and utilization of Tawhidi determined maqasid goods and services and the productive factors for this activity. Technological change, scientific discovery, and analytical discourse around the generalized system framework of Tawhidi methodology based on money and real economy complementary linkages enter the evaluation of the wellbeing criterion.

The portion **M02 → C02 → M12** is caused by withholding of financial capital in the form of idle savings, for example bank savings. As long as this form of saving continues, money is severed from its complementary linkage with the real economy and is driven by substitution into the financial economy. The result is in accordance with the neoclassical theory of opportunity cost and marginal rate of substitution. Savings being opposed to resource mobilization into productive Tawhidi basis of maqasid-codified goods and services permanently denies potential output. Thereby, social wellbeing suffers.

While productive agents are the beneficiaries of the money–real economy linkage, those of the second category form property and property rights of the capitalist masters. Such a capital accumulation gives rise to large financial organizations and speculative capital markets, and portfolio investments as opposed to foreign direct investment and the ensuing overall instability.

Capitalist globalization is entrenched in the above form of dichotomous MCM relationship. The use of money as resource via participatory financial papers replaces interest rates by direct productive and socially rewarding activities. These two activities remain opposed to each other and influences household preferences of choice between the use of money in real economy activities or by substitution into the financial economy. Such choices are institutionalized by banks in society at large.

At the end, in the light of the assumption of scarcity of resources, and thereby systemic competition between alternatives, all of economic theory is entrenched in the marginalist trade-off between the real economy and the savings economy. It is seen in terms of the systemic competition and trade-off between money as mobilized resource by financial instruments into the real economy, against money as withdrawal of resource mobilization by way of bank savings. Thereby, the total wellbeing function represents the marginal rate of substitution between the wellbeing of the corresponding competing sectors for scarce resources. The marginalism here affects all forms of resources, monetary and non-monetary (factors of production). Capitalism cannot get out of this continued form of allocating financial resources between the two competing ends, according to the entrenched neoclassical economic theory of capitalist globalization. By doing so, capitalism perpetually promotes inequality, loss of productivity, and social wellbeing, and causes perpetual instability due to the volatility of the financial sector in response to speculation and the growth of interest-based financial papers. The last point is established by noting that the short-run and long run rates of interest are determined by the term structures of multitudes of interest rates determined by risky contingencies that move randomly in different portfolios.

Towards a Generalized System Money–Real Economy Complementary Model by the Tawhidi Epistemic Methodology

The generalized system understanding of Tawhidi worldview applied to the money and real economy participatory linkages in the light of Tawhidi unity of knowledge is explained by Fig. 8.1 (Choudhury & Hoque, 2003, 2004). In the understanding of good money in the perspectives of Tawhid as law of unity of knowledge, the gold standard proves to be the institutional medium of an ethical currency that can stabilize the economy in terms of prices, output, employment, and a just economic order. These consequences of the gold standard are conveyed in the evaluation of the wellbeing objective criterion with circular causation as organic interrelations between the variables representing maqasid goods in Tawhidi perspective. The Sunnah strongly recommends gold to be adopted as the medium of currency valuation (Alouche, 1994). Imam Ghazali (Karim, undated) pointed out that the currency which does not bear some gold in it cannot be called an Islamic currency. Yet I argue that it is not the gold-bearing currency for circulation that is important. Rather, it is institutional arrangement to hold gold as monetary backing of real assets that is important. It would be acceptable to hold real assets in any form of currency, provided that currency is assigned its value against the gold standard held by the central banks of all Islamic countries.

In Fig. 8.1, the intersection region between Islamic bank and Islamic insurance (takaful) is the starting point of an ethico-economic process that commences with the determination of $\{\theta, \mathbf{x}(\theta)\}$. The arrows spanning outwards indicate the enlarging effect of the *ST*-experience with the region AB widening out and carrying along with this the three institution-market interactive and integrative ethical policy, programme, and organizational variables, namely, EP_1, EP_2, and EP_3.

EP_1 denotes the relationship between central bank and financial institutions, that is, Islamic bank as an example, in the 100 per cent reserve requirement monetary system. In other words, the institutional arrangement then is to have the central bank hold all the reserve of depositors in the Islamic banks. The commercial bank simply matches its clientele with spending outlets, and thereby asks the central bank to supply a quantity of money for which the commercial bank pays a service charge to the central bank called seigniorage.

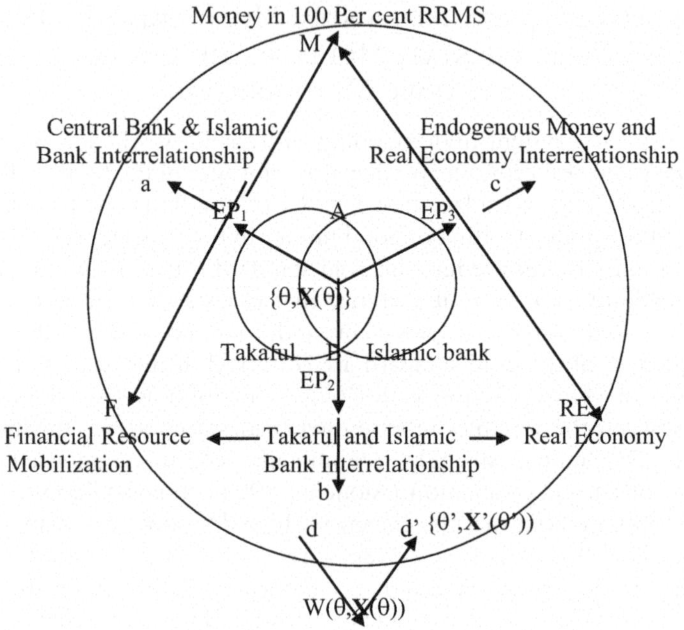

Fig. 8.1 Ethico-economic general equilibrium takaful-economy interaction in 100 per cent reserve requirement monetary system

In the case of the 100 per cent gold dinar-backed currency held by the central bank, the minting of the gold by the central bank is charged to the commercial bank, which the Islamic bank would raise from service charges to clientele.

Besides this, the risk and product diversification is shared between the Islamic bank and the clientele according to their respective profit-sharing ratios. The central bank is primarily an overseer of monetary and economic development and planning authority. The Islamic bank is a clearance house on spending for the central bank's reserve liquidity, which is now managed by the Islamic bank on behalf of the clientele. In this venture, the commercial bank can assume either proportionate risk with other co-operating agents or full risk by lending the total amount. In the latter case, the sharing ratios are determined by adding to the loaned venture capital the imputed production factor costs by the clientele.

EP_1 also includes the financial instruments, such as the principal ones, mudarabah, musharakah, foreign trade, murabaha, bay muajjal, and secondary financing instruments revolving around the principal ones. Above all these, EP_1 also includes the organization of the ST-experience between the central bank, the Islamic bank, and the clientele.

EP_2 denotes the use of the same kinds of financing instruments and ways of realizing risk and product diversification in order to mobilize financial resources from the Islamic banks through non-bank financing institutions. An example of non-bank financial institutions is the Islamic insurance. It acts as the medium of risk-diversification and resource mobilization. Above all these, the organization of the *ST*-experience involving Islamic bank, Takaful, and clientele is foremost in EP_2.

EP_3 denotes activating of market friendly policies and programmes, such as dynamic preference formation (induced by $\{\theta\}$ in the clientele for attaining regimes of life-sustaining goods), rather than 'wants' and conspicuous spending. Through EP_3 it is possible to give a valuation to monetary flows in terms of the returns in the real economy realized by effective financial resource mobilization. Such returns being functions of real market transactions are simply net profit rates. The net profit rates can be expressed as a weighted sum of the profit-sharing ratios. This implies that monetary aggregates are micro-monies specific to markets and clientele in the 100 per cent reserve requirement monetary system (Choudhury, 2004).

The further extending arrows 'abc' denote the evolution of the circular causation interrelations between the $\{\theta, \mathbf{x}(\theta)\}$ variables. At the end of every such circular causation, there is a post-evaluation of the choices and consequences of $\{\theta, \mathbf{x}(\theta)\}$ - variables in the simulated system. Post-evaluation for attained levels of unity of knowledge in complementing possibilities is done by simulation of the wellbeing function $W(\theta, \mathbf{x}(\theta))$ in any given phase of the complete circular causation. The simulation is realized along *ST*-experience. The circular causation arising from such post-evaluation and its continuity is shown by the arrows dd'.

The double-directional arrows along the sides of the money-finance-real economy (MFRE) triangle imply the endogenous nature of the IIE process of circular causation between M, F, and RE. We now derive the Tawhidi rule on phasing out of interest rates. It is not based on an imposed policy of interest abolition as is usually predetermined by governments enforcing shari'ah rule through central bank and Islamic banks. Rather, the Tawhidi rule for effective interest reduction is based on the unifying experience of money-finance-real economy linkages with institutional guidance and enforcement.

Entrepreneurial Capitalism and Money–Real Economy Interrelations

What is the status of entrepreneurial capitalism in the capital accumulation process on which Schumpeter (1961), Hayek (1990), and Weber (1947) wrote their prolific pieces? Entrepreneurial capitalism is a process of innovation premised on continuous innovation by the rise of human creativity. Nothing is static in this process as creative destruction defines the perpetual causation of evolutionary paradigm of innovation and development. Capital formation as a process is realized out of such an evolutionary process of converting money capital into real capital. Yet what is the problem with the theory of entrepreneurial capitalism?

The assumption of scarcity of resources, economic competition, and rational choice behavior continue on to characterize human nature and human ego. Rationalism devolving into economic rationality forms the basis of reasoning (von Mises, 1976). Consequently, the liberal axiom of individual sovereignty remains intrinsic to these Austrian economic doctrines. The combination of the scarcity, competition, and rational choice assumptions once again reverts to maintaining the economic doctrine of marginalist trade-off resulting in substitution instead of complementarities between alternatives. Thereby, savings (as opposed to resource mobilization) and the real economy once again substitute each other in the resource allocation doctrine of the marginalist way of economic reasoning. Hayek (1990) in his *Good Money* brings out this belief on the prevalence of the marginalist school of the neoclassical genre.

Now connecting economics, finance, society, and science in the investigation of the nature of systemic relationship between these, we note that the capitalist globalization paradigm is premised on the liberal epistemology projected in the capitalist world-system (Wallerstein, 1980). The systemic construction takes the following form:

$$\begin{aligned}\text{Liberalism} \rightarrow & \left[\text{Markets, Economy, Science, Society, Technology,}\right.\\ & \left.\text{Global Hegemony}\right](\text{liberalism})\\ & \rightarrow \text{World system as totality of the above system of transnational relations}\\ & = \text{Capitalist world system}\\ & \rightarrow \text{Capitalist globalization.} \qquad (8.7)\end{aligned}$$

Since the above schema is universally inherent in capitalist globalization, therefore the differentiated systemic forms, that is, previously $S(\theta) = \phi$ with $\theta \in$ Rationalism, remain the permanent characteristics of global relations under liberalism. With this kind of systemic permanence of global capitalism, the picture of human future under capitalist globalization is one of conflicting polarity between differentiated parts of the human order. Such a competition picture results in the loss of the social worth of human potential to self-actualize by material growth and moral consciousness. Sztompka (1991, p. 94) explains the role of consciousness in social realization as follows: Consciousness "leads us back to the central claim: what truly exists in society, in the full ontological sense, is the ***unified socio-individual field***, the 'third level of reality' between traditionally conceived levels of totalities and individualities." As well, Nitzan and Bichler (2000, p. 68) write: "From this perspective, the gradual ascent of global capitalism reflects the changing relationship—both contradictory and reinforcing—between state power anchored in sovereignty and capitalist power rooted in ownership."

These conceptual manifestations of the neoclassical and mainstream theories of capitalist globalization negate the possibility for a truly moral and ethical society to emerge, one that can be governed and directed by a law of organic coexistence other than that determined by methodological individualism as the final determiner. Yet there are many who argue that a reformed capitalist globalization with a human face could be possible. Dunning (2004) proposes a top-down approach of institutional guidance, which can be equivalent to the enforcement of the religious beliefs and laws. This would be combined by a bottom-up approach, in which an intensive education and awareness is created at the grassroots levels to understand and adopt the top-down agenda.

Yet at the end of the above explanation, Dunning summarizes the views of other scholars who find that, in the face of current ideas on global ethics, pluralism, and secularism, a post-modern age will not respond to the religious ethical and moral basis of the unitary-type ontology. Consequently, one returns back to the methodological individualism of self-interest in the globalization process even with the clamour of capitalism with a human face. This state then determines the permanent character of liberalism upon which capitalist globalization rests. In the end, the experiment on adorning globalization with a human face disintegrates into a massive contest of individual wills that are then galvanized in the midst of power, hegemony, and predetermined choices. The present political tempo gripping the world and creating stereotypes out of it is sufficient evidence of the catastrophe of a hegemonic collectivization of individual choices and wills under power-driven and culturally differentiated global geopolitics.

Islamic Financing Instruments in the Generalized System Model of Tawhid

Having done the comparative study of mainstream economic reasoning concerning money and real economy, we now return to the methodology of Tawhidi systemic unity of knowledge to explain the extensively participative nature as unity of knowledge induced in financing instruments to bring about participative complementarities between money and real economy. In doing so, we also contrast this Tawhidi epistemological approach of the generalized system model of Tawhid with the fiqhi-type understanding that is prevalent today. Here we invoke again the interactive, evolutionary, and evolutionary intersection of the $S_i(\theta)$ – sets. By the arguments already give above, the presence of unwanted debt, and the overly reliance of Islamic economics and finance to the postulates of neoclassical reasoning, the inextricable presence of shadow rates of interests in the Islamic debt and debt-driven financial system do not purify the financing instruments from such shadow interest rates. This is a methodological issue invoking the need for evolutionary learning towards correcting the prevalent financial instruments from their independent states of uqud, into a generalized system of organic pairing (participatoriness).

We identify the sets $S_i(\theta)$, $\{\theta\}$ being derived from the Tawhidi epistemological premise as follows: i = (1) monetary system, (2) financial system, (3) real economy, (4) resource system, (5) technological and scientific system explaining the analytics of continuous flow of resources. The case (5) involves the formalism of the generalized system model as a mathematical computerized model in non-linear, complex topological space. Thus, the economic, financial, social, and scientific elements are complemented by the induction of $\{\theta\}$ - values by the IIE method of inter-system circular causality, which points to the dynamics of Tawhidi organic unity of being and becoming in the real world.

By the explanations given earlier, the overly contractual (uqud) orientation of independent financing instruments in the prevalent understanding of the shari'ah results in the imminent model of a dialectical type. That is, there is an initial phase of a non-null $S(\theta)$ intersected set. But when this evolves carrying the mainstream and neoclassical behavioural implications of rational choice, the component sets part away; and $S(\theta)$ degenerates into a null set. Methodological individualism and independence returns to the financing instruments, money, and real economy, and between the other sets. This is the real-world scenario of profit-sharing (mudarabah), equity financing (musharakah), cost-plus pricing of real assets under hire-purchase (murabaha), and the secondary instruments, all parting away into their own competing isolation.

A complementary pool of funds comprising all such financing instruments in diversifying the portfolio has not been thought of (Mufeedh Choudhury, 2009). Technological innovation was focused on serving the individualizing ends of independent financing instruments. The principle of participation (complementarity) as the rule derived from the precept of pairing in the Qur'an is lost. Endogenous nature of technological change as opposed to the exogenous one, as found in the case of the traditional contract-centred model of the shari'ah, is the permanent result of knowledge and information generation in the light of Tawhidi unity of knowledge. Indeed, such endogenous knowledge and information flows determine the continuous regeneration of resources across continuums. This is the main reason for annulment of the neoclassical postulate of scarcity of resources and thereby marginal rate of substitution and competition by self-interest.

The Economy According to Tawhidi String Relation

In the end now we establish the pervasively participatory and thus complementary and cooperative nature of the Islamic economy, with which this chapter started on the theme of the Qur'anic precept of pairing, representing Tawhidi unity of being and becoming in the order of knowledge, space, and time dimensions. With this end point we can define an economy denoted by 'E' in reference to the Tawhidi String Relation given by expression (1) and expression (2) in the topological formulation given in expression (9.3) (Choudhury, 1999). See Debreu (1959) for a topological definition of Liberal Economy under perfect competition:

$$E = E\left(\mathbf{R},;\mathbf{p},;\mathbf{q},;\mathbf{P},;\mathbf{T},;W\right)\left[\theta\right] \qquad (8.8)$$

with $\theta \in (\Omega; S)$ being also recursively determined,

Here **R** (resource), **p** (price), **q** (quantity), **P** (policy instruments), **T** (resource mobilizing instruments and technology), and W (wellbeing) are each recursively interrelated according to the unifying relational epistemology of $(\Omega; S)$ through the medium of knowledge flows $\{\theta\}$ so derived and impacting upon these variables and relations. Bold symbols signify vectors. From this broad characterization, specific cases can be derived.

In the case of microeconomics, '*W*' can stand for the objective criterion of knowledge-*simulated* household social wellbeing function with $(\mathbf{R}, \mathbf{p}, \mathbf{q}, \mathbf{P}, \mathbf{T}; W)[\theta]$ representing the corresponding household-specific variables through the recursive learning experience of the household in its Tawhidi experience in unity of knowledge. 'θ' denotes a limiting consensually determined collective and coordinated knowledge-value in decision-making on forms of spending in the good things of life.

In the macroeconomic sense, '*W*' would signify the knowledge-*simulated* social wellbeing function that is aggregated over complex nexus by the IIE process extending over households in concert with the embedded economic, society, and polity circular causation interrelationships. Once again the $(\mathbf{R}, \mathbf{p}, \mathbf{q}, \mathbf{P}, \mathbf{T}; W)[\theta]$ - relations are taken into account over recursively determined knowledge flows derived from the Tawhidi epistemological foundations. Thus, a conscious community in unity of knowledge and the world-systems is established. The Qur'an names such a community as the ummah, the conscious world-nation of Islam.

Conclusion

This chapter as a first of its kind has presented the arguments, explanation, and formalism on the basis the Tawhidi worldview regarding some specific economic and financial issues contra their treatment by independently separated contract-based shari'ah approach. The resulting Tawhidi String Relation (TSR) as formalized identifies two inseparable parts of the purpose and objective of the Tawhidi epistemological core in reference to its emanating moral and ethical constructed world-system. This part arises from the evolutionary learning $\{\theta\}$ - values as simulated over the unification experience of the Qur'anic participative and complementing precept. This attribute is manifested by the paired multiverse nature of Qur'anic world-system. The Tawhidi knowledge flows and their knowledge-induced entities of the world-systems (a'lameen) are ever expanded and strengthened by the increasing comprehension of the Tawhidi ontology leading to epistemology and phenomenology. This chapter has thus expounded the generalized system model of the Tawhidi methodological worldview

Thus, in the TSR characterization, the Tawhidi worldview is not a human law conditioned in the first place by the vagaries of human rationalism, discourse, and interpretation. TSR is not solely based on circumstantial interpretations as of shari'ah rules (fiqh). Rather the necessary and sufficient condition for the derivation of Tawhidi rules (ahkam) of unity of knowledge is the primacy and direct reference to its primal ontology and epistemology in the total tasbih-shura (*ST*) conscious process of knowledge formation and its ontological construction of the knowledge-induced world-systems in knowledge, space, and time dimensions.

The emanating dynamics of the Tawhidi evolutionary learning experience in the multiverse generalized system framework emulating the Tawhidi epistemic worldview are represented in a number of problems of economics and finance interrelating with society and science.

How therefore must Muslims enter a post-modern age of socio-scientific thinking? This chapter has rejected the existing overly legal contractual aspects of the shari'ah governing independent piecemeal attention to issues of economics, finance ,and society and excluded socio-scientific inclusion. Institutional developments on the basis of the widest understanding of the Tawhidi worldview have remained unknown. This state of the shari'ah discourse is to be abandoned for a holistic understanding of the multiverse by the overarching generalized system formalism premised on the Tawhidi episteme. The traditionalism of Islamic jurisprudence (fiqi Taqlid) and Muslim subservience to liberalism has protracted the models of Islamic economics, finance, and society thus far. Yet Islam and liberalism are proved to be unmixable (Murden, 2002).

Towards attaining this objective end, the following broad recommendations are proposed here. But one must note that these are simply suggested as points for further discourse. No detailed analysis is given here. The reader may turn elsewhere for such functional details that bring the Tawhidi methodology and worldview into empirical and applied scientific constructs.

1. The institutions of Islamic thought and practice ought to consider implementation of the Tawhidi methodological worldview in the construction, formalism, and application of the framework of Islamic economics, finance, science, and society within a generalized system model of unity of knowledge (i.e. complementarities and participation).

2. To achieve the implementation of (1), open and specialized forums for the understanding and application of the Tawhidi methodology and worldview in both human resource and institutional development ought to be promoted in every possible and pertinent forum. In this phase of Tawhidi knowledge dissemination, the most modern and sophisticated of methods ought to be applied in order to galvanize the Tawhidi methodology and worldview into quantitative world-systems dynamics.
3. Specifically, the monetary and financial system of the Muslim world ought to be transformed into a money–real economy complementary and strong circular causation linkage system. Likewise, the debt-based financial instruments of Islamic financing ought to replace the state of debt instruments presently existing. This will liberate all forms of productive and maqasid-based resources of Tawhid as law for the wellbeing of the ummah transformation. It will also yield a new way of morally centred socio-scientific thinking for academia and practitioners.
4. In achieving all such transformation, the maqasid-based financial and developmental resources in the light of Tawhid as law ought to be mobilized by the full gamut of Tawhidi methodological thinking in the multiverse socio-scientific world-system.

References

Alouche, A. (1994). *Mamluk Economics: A Study and Translation of Al-Maqrizi's Ighatah.* Salt Lake City, Utah: University of Utah Press; specially the chapter, "currency".

Asad, M. (1987). *This Law of Ours.* Gibraltar: Dar al-Andalus.

Buchanan, J. M. (1999). The Domain of Constitutional Economics. In *The Collected Works of James M. Buchanan, Vol. 1.* Indianapolis, IN: Liberty Fund.

Choudhury, M. A. (1991, July). The Concept of 'A New World Order' in Relation to the Gulf Crisis. *The Universal Message.*

Choudhury, M. A. (1999). Definition of Economy. In *Comparative Economic Theory, Occidental and Islamic Perspectives* (p. 71). Norwell, MA: Kluwer Academic.

Choudhury, M. A. (2004). Micro-Money and Real Economic Relationship in the 100 Per Cent Reserve Requirement Monetary System. *Review of Islamic Economics, 13*, 1.

Choudhury, M. A. (2009). *Money, Finance and the Real Economy in Islamic Banking and Finance, Perspectives from the Maqasid as-Shari'ah.* MSc dissertation, Department of Economics, Banking and Finance Program, University of Stirling, Scotland.

Choudhury, M. A. (2013). Complexity and Endogeneity in Economic Modeling. *Kybernetes, International Journal of Systems, Cybernetics, and Management Sciences, 42*, 2.

Choudhury, M. A., & Hoque, M. Z. (2003). Islamic Finance: A Western Perspective—Revisited. *International Journal of Islamic Financial Services, 5*(1) electronic version.

Choudhury, M. A., & Hoque, M. Z. (2004). An Ethico-Economic General Equilibrium Study of Linkages Between Money, Finance and Real Economy Using Islamic Insurance. In *An Advanced Exposition of Islamic Economics and Finance* (pp. 221–248). Lewiston, NY: The Edwin Mellen Press.

Debreu, G. (1959). *Theory of Value, an Axiomatic Analysis of Economic Equilibrium*. New York, NY: John Wiley.

Dunning, J. H. (2004). The Moral Imperatives of Global Capitalism: An Overview. In J. H. Dunning (Ed.), *Making Globalization Good* (pp. 11–40). Oxford: Oxford University Press.

Hayek, F. A. (1990) (reprint). The Use of Knowledge in Society. In M. C. Spechler (Ed.), *Perspectives in Economic Thought* (pp. 183–200). New York, NY: McGraw-Hill.

Heilbroner, R. L. (1985). The Ideology of Capital. In *The Nature and Logic of Capitalism*. New York, NY: W.W. Norton.

Karim, F. (undated). *Imam Ghazzali's Ihya Ulum-Id-Din*. Lahore: Shah Muhammad Ashraf.

Mundell, R. A. (2000, November 8). Presentation at the Economic Forum on "One World, One Currency: Destination or Delusion?". Washington, DC: International Monetary Fund. Retrieved from http://www.imf.org/external/np/tr/2000/tr001108.htm.

Murden, S. W. (2002). Islam and the Liberal Idea. In *Islam, the Middle East and the New Global Hegemony* (pp. 155–184). London: Lynne Rienner.

Nitzan, J., & Bichler, S. (2000). Capital Accumulation: Breaking the Dualism of 'Economics' and 'Politics. In R. Palan (Ed.), *Global Political Economy*. London: Routledge.

Sardar, Z. (1984, December). Islamisation of Knowledge or the Westernisation of Islam? *Inquiry, 1*, 7.

Sardar, Z. (1988). *Islamic Futures, the Shape of Ideas to Come*. Petaling Jaya: Pelanduk Publications.

Schumpeter, J. A. (1961). *The Theory of Economic Development*. Chapter II (R. Opie, Trans.) Cambridge, MA: Harvard University Press.

Sztompka, P. (1991). The Model of Social Becoming. In *Society in Action, the Theory of Social Becoming*. Chicago, IL: University of Chicago Press.

Underhill, G. R. D. (2003). Conceptualizing the Changing Global Order. In R. Stubbs & G. R. D. Underhill (Eds.), *Political Economy and the Changing Global Order* (pp. 3–23). London: Oxford University Press.

Ventelou, B. trans., & edited with Introduction by Nowell, G.P. (2005). "Keynesian Economics and the American Polity", being the "Introduction" by G. P. Nowell. *Millennial Keynes* (pp. 3–37). Armonk, New York: M.E. Sharpe.

Von Mises, L. (1976). The Human Mind. In *The Ultimate Foundation of Economic Science* (pp. 11–33). Kansas City: Sheed Andrews and McMeel, Inc.

Wallerstein, I. (1980). *The Modern World-System.* New York, NY: Academic Press.

Weber, M. (1947). *Economy and Society.* New York, NY: Oxford University Press.

CHAPTER 9

A Mathematical Model of Symbiotic Aggregation and Participatory Dynamics in Tawhidi Epistemological: Worldview of Unity of Knowledge

Introduction

The Tawhidi methodological worldview bestows its critical foundation on systemic consilience, which itself explains the functional ontology of unity in diversity by the organic relationship between being. The collectivities of such beings represent the nature of organismic aggregation. Thereby, the nature of aggregation of functions in their mathematical form in all of Islamic methodological worldview of unity of knowledge is fundamental to understand the meaning and relational functioning of participation. The participative nature of everything is of the relational category. Their functional expressions are of the linear type avoiding complexity. Non-linearity pronounces a substantively defined nature by its essential property of feedback inter-causality between diversely interactive, integrating,

Professor Dr. Masudul Alam Choudhury and Mr. Mahfuzul Alam Taifur, doctoral program student in the Department of Shari'ah and Economics, Academy of Islamic Studies, University of Malaya.
The chapter was supervised in writing and developing ideas by Professor Dr Masudul Alam Choudhury, Postgraduate Program in Islamic Economics and Finance, Faculty of Economics, Trisakti University, Jakarta, Indonesia; and Visiting Professor, Social Economy Center, Ontario Institute for Studies in Education, University of Toronto. masudc60@yahoo.ca

M. A. Choudhury, *Tawhid and Shari'ah*,
https://doi.org/10.1007/978-3-030-49087-4_9

and evolutionary (IIE) learning processes of variables and their monotonically transformed functions.

All of these mathematical modelling features together remain intrinsic to the dynamic nature of participation by way of complexity of aggregation in the complementary implication of unity of knowledge by way of interrelations. Such interrelations in the scope of participation by mutuality of inter-causal learning as they represent the grand and particular cases of diversity in multitude of systems provide the essential nature of morality and ethics. Such inter-causal relations between the good things of life explain the Qur'anic meaning of Tawhid in its methodological action in the order and scheme of everything. Yet the same methodological model, though in aggregation by differentiation between variables and their functions, also explain social differentiation and thereby social exclusion.

The presence of continuous interaction, integration, and process-based evolution of affected forms altogether explain the unity of knowledge by complementarities as the sure sign of social inclusion and thus the design of the symbiotic universe with embedding of morality with materiality. On the other hand, the disintegration by lateral addition of independently isolated relations is the nature of aggregation in the state of social differentiation as opposed to moral embedding of material forms. The dynamics of aggregation is thus shown in two distinct forms. They are by complexity caused by richness of embedded complementary relations between variables of the good things of life (halal at-tayyaba) or by separation of independent forms. The consequential form by cause and effect of the first kind will be termed as 'wellbeing' (maslaha) in this chapter. The imminent type of the second kind is of the utilitarian form referred to as welfare function in welfare economics.

Objective

The above form of distinction in the complex and linear forms of mathematical functions defining the relationship between variables gives important organizational inferences of inter-causal systems contrary to independently distributed linear forms. Thus, the central role of participation and its complex aggregation form different analytical formalism in a substantively understood of participation between complementary possibilities contrary to marginalist opposites. In the context of such relational implications between aggregation and participation and the imminent formal models, important abstracto-empirical derivations arise to help in policy and institutional recommendations. The study of such a kind is the objective of this chapter.

Formalizing Aggregation as Symbiotic Participation in Islamic Methodological Worldview

We characterize the analytical difference between the above-mentioned two forms of aggregations by their opposite forms of complementarities and methodological independence that differentiates between the definition of participation as organic symbiosis of unity of knowledge and marginalism contra participation in differentiated aggregation. Because of the vast implication that such aggregation concepts convey, the interrelating principle of participation in 'everything' in Islamic ontological and epistemological methodological worldview of unity of knowledge conveys a universal meaning of symbiosis. There are deep implications of the properties relating to the nature of aggregation and participation in a morally inclusive methodological worldview, as of the Islamic reflection of inter-variables systemic causality between the recommended things. Such properties and implications lead into the characterization of the empirical form of the conceptual worldview of complementarities in the model as the sure sign of Tawhidi unity of knowledge. From the ensuing empirical results and implications of the analytical properties of the various elements of the mathematical model, policies and institutional reconstructive forms can be derived for a sustainable socially inclusive future. These characteristics and features of the Tawhidi worldview in respect of aggregation and participation are now formalized below.

Brief Review of the Literature in Complex Aggregations

Interest in the area of aggregation in complex fields has drawn a growing interest in recent times. The emergence of complexity theory has premised mathematical formalism in non-linear functions (Bertuglia & Vaio, 2005; Choudhury, 2013). Since a long time now, philosophical constructs had engaged themselves on how scientific reasoning is viewed in socio-scientific worldview by rationalism (Kant, I. (trans. Friedrich, C.J., ed.), 1949). A semblance of morality and ethics within the theory of justice and institutional reconstruction was recently explained by Rawls (1971).

Yet such earlier socio-scientific works were critically viewed in the context of what truly is the ultimate reality in the formulation of socio-scientific reasoning. In this regard Bhaskar (2002) gave his theory of critical realism. Carnap (1966) criticized the dichotomous yet a desired worldview of the moral imperative by Kant. A similar kind of rejection of the a priori concept of moral and ethical premise is voiced in the concept of pseudo-science and demarcation by Popper (2004). The lack of moral explanation in economic theory has been questioned by Sen (1999) in his definition of 'wellbeing' distinctly apart from the neoclassical idea of 'welfare'.

In the whole history of dialectical reasoning that delineates socio-scientific thought, the epistemological premise of unity of knowledge has remained distanced. A recent example of this failure is found in the demise of superstring theory in theoretical physics (Callender, 2010), and by the questioning launched by the heterodox economic school as by Lawson (2003). In the midst of the pending search for such a fresh epistemic premise of unity of knowledge, the Tawhidi methodological worldview gives a fresh and universal and unique possibility for the socio-scientific foundation of abstracto-empirical modelling within unity of knowledge and its participatory implications and symbiotic nature of aggregation.

Mathematical Formalism of Aggregation Qua Participation

Participation

First, a general meaning of participation is provided in mathematical form. Such a definition will apply to all kinds of specific-problems and systemic forms of interrelationship with the principle of universal complementarities and their reconstructive formation.

Let $\mathbf{x}(\boldsymbol{\theta})$ denote the vector of variables induced by parameters of unity of knowledge, 'θ':

$$\mathbf{x}(\theta) = \{x_1, x_2, \ldots, x_n\}[\theta], \text{such that}, \{\mathbf{x}_i\} \cap \{\mathbf{x}_j\} \neq \phi, i, j = 1, 2, \ldots, n; i \neq j. \; \{\mathbf{x}_i\} \cup \{\mathbf{x}_j\} = \mathbf{x}(\theta). \tag{9.1}$$

in the inter-variable relational sense.

By the properties of well-definition, continuity, and thus the topological property of the above-mentioned sets by the existence of learning denoted by the knowledge-induction parameter, {θ}, a monotonic function, say, f(.), is topologically well defined on the $\mathbf{x}(\boldsymbol{\theta})^4$ (Maddox, 1970). We can thereby derive,

$$f(\{\mathbf{x}_i\} \cap \{\mathbf{x}_j\})[\theta] \neq \phi; f(\{\mathbf{x}_i\} \cup \{\mathbf{x}_j\})[\theta] \\ = f(\{\mathbf{x}_i\}) \cup f\{\mathbf{x}_j\} - f(\{\mathbf{x}_i\} \cap \{\mathbf{x}_j\}. \tag{9.2}$$

This is the case of interdependent relational functions as shown here. All positive monotonic topological functional transformations of f(.) follow the same rule as mentioned here (Henderson & Quandt, 1971). Thus, these topological properties of evolutionary functions in θ - parameters by monotonic functional transformation exist for such transformations indefinitely, unless 'de-knowledge'[1]-induced variables intervene to break up the topological properties and continuity of functions in θ - induction.

Nature of Aggregation of Variables and Their Topological Functional Transformation

Aggregation: Welfare Criterion of Mainstream Economics

In the case of relational independence between the variables of $\mathbf{x}(\boldsymbol{\theta})$ - vector expression (9.2) becomes,[2]

$$f(\{\mathbf{x}_i\} \cup \{\mathbf{x}_j\})[\theta] = f(\{\mathbf{x}_i\}) \cup f\{\mathbf{x}_j\} \tag{9.3}$$

[1] θ - parameters signify unity of knowledge as derived from the Tawhidi primal ontological foundation which exists in the case. De-knowledge is also shown by ranks of differentiation between variables that are contrary to unity of knowledge, that is, complementarities between the good things of life, that is, signifying inter-variable marginalism. There is no need for constructing a differently new set of de-knowledge parameters other than the knowledge parameters. De-knowledge as inter-variable differentiation, marginalism, and social exclusion will be indicated by the negative estimated coefficients between the recommended variables, while avoiding the morally contrary ones as dictated by Tawhid as the law of unity of knowledge.

[2] A form of this functional relationship is the probability of independently occurring variables. Thereby, $f(\{\mathbf{x}_i\}) \cap f\{\mathbf{x}_j\} = \phi$

According to the concept of optimal objective criterion in economics, the variables or their distinguished groups of the optimal function form marginalist combinations. This implies, $df(\mathbf{x}(\theta)) = 0$. Examples of $f(\{\mathbf{x}_i\} \cup \{\mathbf{x}_j\})[\theta] = f(\{\mathbf{x}_i\}) \cup f\{\mathbf{x}_j\}$ are the welfare function with interpersonal utility as variable components and production function with capital and labour as marginalist opposites by their marginal rates of substitution.

The aggregate nature of dissociated variables and their independently distributed functions as of expression (9.3) form the utilitarian welfare function (Harsanyi, 1955; Hammond, 1989). By the very properties underlying the functions in expression (9.3), the property of complementarities between variables and their topologically transformed functions is eschewed. Marginalism prevails always and in everything. The generalized inference from these linearity properties of additively aggregated variables and their functions is this: Linear structure of independently distributed variables and their topologically transformed functions with constant coefficients either in estimated or simulated form (evaluation) imply the persistence of marginalism and permanent avoidance of complementarities (Simon, 1987).

Wellbeing Criterion of Tawhidi Economics with Pervasive Complementarities

The following is the general formula of composite variables and their functional in complex interactive, integrative, and evolutionary (IIE) learning processes:

Interaction

In reference to the vector of variables, $\mathbf{x}(\boldsymbol{\theta})$, the property of interaction between the elements of the variables is explained by the endogenous inter-variable circular causality explained by the system of interrelations,

$$x_i(\theta) = f_i(\theta, \mathbf{x}_j(\theta)), i, j - 1, 2, \ldots, n; i \neq j., \qquad (9.4)$$

with the elements of $\mathbf{x}(\boldsymbol{\theta})$ being desired to be complements by estimation or simulated by continuously dynamic estimated coefficients by their Bayesian distribution (Hogg & Craig, 1965).[3]

[3] Continuous coefficients are generated by their conditional probability distribution.

Integration
Integrative learning is determined by one complete evaluation (estimation and simulation) of the system of equations given by the solution of the expression (9.4) for all circular-causation equations. Thus, integration of the interactive learning system by evaluation forms a learning solution of the endogenously complementary circular causation relations as a computational general equilibrium result with θ - learning parameters.

Interaction Leading to Integration in Continuous Domains: Intra-System Unity of Knowledge
Interaction and integration for any given stage of evaluation of the system of equations lead into that stage-specific evaluation of the objective learning function. We refer to such a function as the wellbeing function. The learning wellbeing function for any stage of interaction and integration represents the degree of complementarities attained by the interacting and integrating sets of variables and their monotonic transformed functions. These are the sequences of estimated and simulated results of the wellbeing criterion within a given stage. The conceptual form of the wellbeing function in the first instance of the objective learning function in unity of knowledge, subject to the system of circular causation equations, leads into the quantitative form of the same wellbeing function, as a scalar transformation in polynomials in θ - parameter. The simple form of such a quantitative empirical form of the conceptual wellbeing function is denoted by,

$$\vartheta = F(\mathbf{x}(\theta)) \tag{9.5}$$

The evaluation of the wellbeing function, subject to the system of circular causation equations for any given stage, is referred to as the intra-system learning process. Its complete intra-stage quantitative model is given as follows:

$$\text{Evaluate } f(\{\mathbf{x}_i\} \cap \{\mathbf{x}_j\})[\theta] = W(\mathbf{x}(\theta)) \text{ as the conceptual form} \tag{9.6}$$

changes into the quantitative empirical form,

$$\textit{Interactions}, x_i(\theta) = f_i(\theta, \mathbf{x}_j(\theta)), i, j - 1, 2, \ldots, n; i \neq j. \tag{9.7}$$

leading into,

integration by evaluating (estimation and simulation of coefficients) this expression (9.7),

and

quantitative empirical form of the reduced form of the conceptual wellbeing function,

$$\vartheta(\theta) = F(\mathbf{x}(\theta)) \tag{9.8}$$

Expression (9.8) is the evaluated form of the θ - function in quantitative form with ϑ assuming its values over the averaged θ - values.

Expressions (9.6)–(9.8) form the intra-system evaluation of wellbeing function as the objective of the learning system according to the precept of unity of knowledge.

Learning Process

The completion of expressions (9.6)–(9.8) on any intra-system stage marks the attainment of a learning process that is centrally governed by knowledge, which in the discursive way as explained above is the unity of knowledge signified by θ - parameter for that particular intra-stage marked by the completion of interactions leading into integration and the accomplishment of the abstracto-empirical moral-material unification of inter-variables in the wellbeing function. We therefore write,

$$\text{Process}, P(\theta) = \{\{\theta\} \in \text{Primal Ontology of unity of knowledge}^{8}, \\ \mathbf{x}(\theta), W(\theta) : t(\theta)\} \tag{9.9}$$

Evolution

Evolution marks the new, that is, inter-system phase of [t]ime-repeated process by learning in terms of complementarities of $\{\mathbf{x}(\boldsymbol{\theta}), W(\theta)\}$ and its monotonic functional transformation, say $P(\mathbf{x}(\boldsymbol{\theta}), W(\theta)$ over time. We denote the evolutionary property of the process with dynamic change into inter-system stage by,

$$(d/dt)(dP(\theta)/d\theta) = (d/dt)\sum_i [\partial P(\ldots)/\partial x_i(\theta)].[dx_i/d\theta] \\ +(d/dt)\sum_i [\partial P(\ldots)/\partial \vartheta(\theta)].[d\vartheta/d\theta] \tag{9.10}$$

W takes the approximate form given by expression (9.8).

$$d\vartheta / d\theta = \sum_{i-1}^{n} \left[\partial F(.)/\partial x_i(\theta)\right].\left[dx_i(\theta)/d\theta\right]$$

Evolution follows the abstracto-empirical evaluation of wellbeing, subject to circular causation relations at the end of a process, evolving thus into the continuous onward processes over {knowledge, space, time} [θ] - dimensions.

It is to be noted that each process over time experiences its own evaluation of expression (9.9) by the expressions (9.6)–(9.8). Thereby the comparative intertemporal values of wellbeing can be charted.

Note the cross-sectional double-sided arrows. They explain the evolutionary intra-system and inter-system evolutionary learning processes by participation (complementarities) continuously across {knowledge, space, time}[θ] - dimensions, which is everywhere along the continuous knowledge-induced world-system. Multiple evolutionary learning trajectories fill the entirety of the {knowledge, space, time}[θ] - dimensions. The intra-system stages of interaction leading to integration are represented by the evolutionary trajectories of Fig. 9.1.

Example from Islamic Financing Instruments

We take the example of [M_1]udarabah, [M_2]usharakah, and [M_3]urabaha as the three well-known Islamic participatory financing instruments. We relate their IIE learning processes and continuous evolutionary learning in Fig. 9.1. The two cases of aggregation in the non-participatory utilitarian case and the Tawhidi participatory case are shown. The overarching circle for the two different natures of interrelations between M_1, M_2, and M_3 denotes the portfolio of the independently distributed case in utilitarianism,[4] and the portfolio of the IIE nature in the Tawhidi context.

In the utilitarian case of the financial portfolio, 'welfare' criterion of the portfolio is denoted by, say, U('deknowledge') = $a_1M_1 + a_2.M_2 + a_3.M_3$; 'a's' are constant coefficients. In the IIE-symbiotic case of learning in unity of knowledge, Tawhidi wellbeing is denoted by, $W(\theta) = \prod_{i=1}^{n} M_i(\theta)^{ai(\theta)}$. The important properties of the IIE learning wellbeing index is firstly, the

[4] Paradoxically, this is the nature of Islamic financing instruments in their independently distributed contractual relations of 'aqd'.

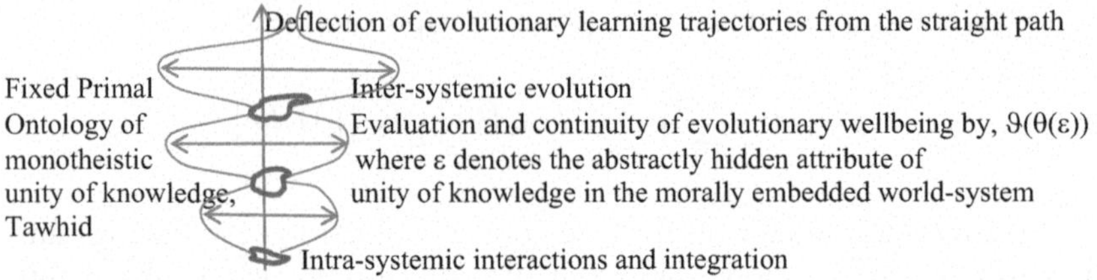

Fig. 9.1 Representation of intra-system and inter-system learning processes of unity of knowledge and the morally embedded world-system

Table 9.1 Bangladesh annual rates of growth and assignment of wellbeing (ethical) values, 1971–2009

Year	*Population*		θ_1	θ_2	θ	*Average*
	Rate of growth of GDP per capita (USD) (%) (g_1)	*Rate of growth of GDP per capita (%)* (g_2)				
1972	2.71	89.72	14.23	9.29	10.00	9.645
1973	2.72	89.98	0.29	9.28	2.29	5.785
1974	2.73	104.67	16.32	9.27	10.00	9.635
1975	2.73	106.96	2.18	9.27	4.18	6.725
1976	2.73	116.34	8.77	9.27	10.00	0.635
1977	2.72	119.28	2.53	9.28	4.53	6.905
1978	2.72	148.82	24.77	9.28	10.00	9.640
1979	2.72	167.69	12.68	9.28	10.00	9.640
1980	2.71	185.06	10.36	9.29	10.00	9.645
1981	2.71	182.75	−1.25	9.29	0.00	4.645
1982	2.69	161.96	−11.38	9.31	0.00	4.655
1983	2.66	161.01	−0.59	9.34	0.00	4.670
1984	2.61	183.93	14.24	9.39	10.00	4.695
1985	2.54	186.12	1.19	9.46	3.19	6.325
1986	2.47	191.59	2.94	9.53	4.94	7.235
1987	2.40	212.58	10.96	9.60	10.00	9.800
1988	2.34	224.36	5.54	9.66	7.54	8.600
1989	2.28	238.28	6.20	9.72	8.2	8.960
1990	2.22	243.33	2.12	9.78	4.12	6.950
1991	2.16	254.52	4.60	9.84	6.6	8.220
1992	2.10	254.45	−0.03	9.90	0.00	4.950
1993	2.06	257.41	1.16	9.94	3.16	6.550
1994	2.02	268.15	4.17	9.98	6.17	8.075
1995	1.99	295.63	10.24	2.00	10.00	6.000
1996	1.97	304.69	3.06	9.97	5.06	7.515

(*continued*)

Table 9.1 (continued)

Year	*Population*		θ_1	θ_2	θ	*Average*
	Rate of growth of GDP per capita (USD) (%) (g_1)	*Rate of growth of GDP per capita (%)* (g_2)				
1997	1.94	309.20	1.48	9.94	3.48	6.710
1998	1.91	314.51	1.72	9.91	3.72	6.815
1999	1.87	323.78	2.94	9.87	4.94	7.405
2000	1.85	323.01	−0.24	9.85	0.00	4.925
2001	1.79	317.07	−1.84	9.79	0.00	4.895
2002	1.75	323.70	2.09	9.75	4.09	6.920
2003	1.70	348.60	7.69	9.70	9.69	9.695
2004	1.65	371.20	6.48	9.65	8.48	9.065
2005	1.59	376.35	1.39	9.59	3.39	6.490
2006	1.53	387.93	3.08	9.53	5.08	7.305
2007	1.47	434.85	12.09	9.47	10.00	9.735
2008	1.42	493.75	13.54	9.19	10.00	9.595
Average	**2.23** ($\theta *_1 = 10$ for avg. rate = 2.00%)		**4.31** ($\theta *_2 = 10$ for avg. rate = 8%)			

Source: Database, Statistical, Economic and Social Research and Training Center for Islamic Countries, Ankara, Turkey

complementary (participatory) relationship between the M_i-variables, i = 1,2,3. Secondly, there ought to be sustainability of change in the participatory financing instruments, not one against the rest.[5]

An Empirical Work on the Circular Causation Estimation

Statistical Results: Population and Economic Growth Relationships in the Circular Causation Model

A complex systemic relationship governing the population growth debate ought to lead the way institutionally and should be academically evaluated within a robust ethical index of human wellbeing. Such a model ought

[5] $W(\theta) = \prod_{i=1}^{n} M_i(\theta)^{ai(\theta)}$ results in the form of rates of change, $g_w = \sum_{i=1}^{n}[a_i.\ g_i(M_i)][\theta]$. The inter-causality between M_i, and thereby between $g_i(M_i)$, signified by positive complementary signs of $a_i(\theta)$, i=1,2,3.

then to unravel the selection of those variables that complement together in order to heighten the human wellbeing index. The empirical work in respect of the policy guidance based on the Tawhidi IIE learning kind of worldview was carried out.

The policy implications of our result together with the supporting arguments expounded in this chapter on the epistemological basis of sustainability in the population growth development debate, as an example, are the following: Population-growth dynamics do not follow a marginalist substitution agenda as otherwise assumed by the neoclassical economic argument, which is here critically examined in reference to the immiserization growth theory and contrarily rejected by alternative development theories. Consequently, there ought not to be such policies formulated, and thereby large national resources spent by emphasizing the trade-off argument between population and economic growth.

These values are targeted because of development sustainability and realism around these points.

Formulas applied for estimating θ_1 and θ_2 values:

1. $\theta_1 = (g_1 - 2) \pm 10$, subject to an inverted curve-values for g_1 on either side of the 2% optimal value of g_1.
2. $\theta_2 = (g_2 - 8) \pm 10$, subject to the values $\theta_2 = 10$ for $g_2 > 8$; and $\theta_2 = 0$ for $g_2 < 0$.

Summary Results for Bangladesh Data

Model 1: Regression Analysis: g_2 and θ on g_1

$$g_1 = 2.64 + 0.0206\, g_2 - 0.0730\, \theta$$
$$t = (9.59)\ (1.57)\ (-1.76)$$
$$R^2 = 9.6\%,\ SE = 0.43.$$

g_2 is statistically significant at 10% level.
θ is statistically significant at 5% level.
Durbin-Watson statistic = 0.226865
D-W statistic indicates high positive autocorrelation.

Model 2: Regression Analysis: g_1 and θ on g_2

$$g_2 = -15.5 + 3.29\, g_1 + 1.87\, \theta$$
$$t = (-2.52)\ (1.57)\ (4.22)$$
$$R^2 = 35.3\%,\ SE = 5.45$$

g_1 is statistically significant at 10% level.
θ is statistically significant at less than 1% level.
Durbin-Watson statistic = 2.22584
D-W statistic indicates almost no autocorrelation.

Model 3: Regression Analysis: g_1 and g_2 on θ

$$\theta = 8.75 - 1.14\, g_1 + 0.184\, g_2$$
$$t = (5.98)\ (1.76)\ (4.22)$$
$$R^2 = 36.4\%,\ SE = 1.71$$

g_1 is statistically significant at 5% level.
g_2 is statistically significant at less than 1% level.
Durbin-Watson statistic = 2.01858
D-W statistic indicates no autocorrelation.

Conclusion: Up-Winding

Derivation of the wider nature of Islamic socially inclusive and socio-economic development by participatory and symbiotic learning system has been uniquely exemplified by a process-based learning organism that pairs with the good things of life and avoids the bad things. The meaning so conveyed by the symbiotic nature of participation and aggregation of the distinctive model implies sustainable continuums of interrelations. This conveys the underlying idea of the ontological-epistemic nature of unity of knowledge. Consequently, in the absence of the wider field of valuation of the participatory Islamic system of inter-causal relations as a socially inclusive, developmental, and sustainable idea, the Islamic understanding of unity of knowledge has been missed out as the holistic interactive, integrative, and evolutionary (IIE) learning worldview as methodology arising from the Tawhidi ontological foundation.

References

Bertuglia, C. S., & Vaio, F. (2005). Dynamical Systems and the Phase Space. In *Nonlinearity, Chaos & Complexity, the Dynamics of Natural and Social Systems* (pp. 49–70). Oxford, UK: Oxford University Press.

Bhaskar, R. (2002). *Reflections on Meta-Reality, Transcendence, Emancipation, and Everyday Life*. New Delhi, India: Sage Publications.

Callender, C. (2010, September). Stephen Hawking Says There's No Theory of Everything. Cultural Lab (Internet).

Carnap, R. (1966). Kant's Synthetic A priori. In M. Gardner (Ed.), *Philosophical Foundations of Physics*. New York, NY: Basic Books, Inc..

Choudhury, M. A. (2013). Complexity & Endogeneity in Economic Modeling. *Kybernetes, 42*(2), 226–240.

Hammond, P. J. (1989). On Reconciling Arrow's Theory of Social Choice with Harsanyi's Fundamental Utilitarianism. In G. R. Feiwel (Ed.), *Arrow and the Foundation of the Theory of Economic Policy* (pp. 179–221). London: Macmillan.

Harsanyi, J. C. (1955). Cardinal Welfare, Individualistic Ethics, and Interpersonal Comparisons of Utility. *Journal of Political Economy, 63*, 309–321.

Henderson, J. M., & Quandt, R. E. (1971). *Microeconomic Theory* (pp. 19–20). New York: McGraw-Hill.

Hogg, R. V., & Craig, A. T. (1965). *Introduction to Mathematical Statistics*. New York, NY: Macmillan Co.

Kant, I. (trans. Friedrich, C.J. ed.). (1949). Critique of Pure Reason. In *The Philosophy of Kant*. New York: The Modern Library.

Lawson, T. (2003). An Evolutionary Economics? In *Reorienting Economics* (pp. 110–140). London: Routledge.

Maddox, I. J. (1970). *Elements of Functional Analysis*. Cambridge, Eng: Cambridge University Press.

Popper, K. (2004). *The Logic of Scientific Discovery*. London, Eng: Routledge.

Rawls, J. (1971). *A Theory of Justice*. Boston, MA: Belknap Press of Harvard University Press.

Sen, A. (1999). Functioning and Well-being. In *Commodities and Capabilities* (pp. 17–21). New Delhi, India: Oxford University Press.

Simon, H. (1987). Decision Making and Organizational Design. In D. S. Pugh (Ed.), *Organizational Theory* (pp. 202–223). Harmondsworth, Middlesex, Eng: Penguin Books.

CHAPTER 10

Conscious Technology and Technological Change in the Light of Tawhid: Artificial Intelligence, FINTECH, and Other

Introduction and Background

Technology in its overarching meaning embodies moral and ethical values that are embedded in material values in an endogenous design of construction, measurement, and policy programme dynamics of technological choice. This formalism gives a phenomenological way of understanding and implementing the conscious nature of technology. Such an embedded nature of consciously integral machines is found in the present days' applications of neural formalism and artificial intelligence (Kaku, 2015). Kaku explains the idea of phenomenology of conscious technology in the following words[1]: "Consciousness is the process of creating a model of the world using multiple feedback loops in various parameters (e.g. in temperatures, space, time, and in relation to others), in order to accomplish a goal (e.g. find mates, food, shelter)." Kaku calls the above definition of phenomenological consciousness as the 'space-time theory of consciousness'.

In this chapter, consciousness (phenomenology) will be referred to in terms of its spanning of values over knowledge, space, and time dimensions in which the fruition between material and moral values occurs. Thus, in the general case of technology and technological change as a medium of development transformation, it is necessary to incorporate the

[1] Kaku, M. (2015). "Consciousness—a physicist's viewpoint", in his *The Future of the Mind*, Chapter 2, Anchor Book, New York, NY.

M. A. Choudhury, *Tawhid and Shari'ah*,
https://doi.org/10.1007/978-3-030-49087-4_10

moral values with the material values to project the goal of wellbeing. Such a wellbeing criterion is seen as conscious moral-material development transformation when these values are endogenously interrelated in accepting positive inter-variable relations and rejecting the marginal rate of substitution or differentiation and trade-off by way of social Darwinism (negative complementarities).[2]

The objective criterion of the wellbeing function is meant to evaluate degrees of positive complementarities in terms of the phenomenology of unity of knowledge. Such positive complementarities result from endogenous inter-variable relations. Otherwise, any given degree of differentiation between the variables could arise. The choice of positive complementarities arising from the organic unity of relations between selected indicators marks the morally inclusive choice of technology, technological development, technological change, and thereby sustainability along the development regime.

Objective

The objective of this chapter is to present the idea and its evaluation in technological choice as a consciously learning way and the possibility of its transformation into a moral-material embedded medium of social inclusiveness. Here technology in general as an adaptive social artefact of inclusive development or the specific example of FINTECH can be used to attain the goals of social wellbeing that is manifest in inter-variable unity of knowledge and the resulting circular causation equations that arise from the endogenous nature of the selected moral and material variables. The central implication made is that the overarching domain of human motivation and the acts of imparting wellbeing through technology and technological change are to be combined in the worldview of moral inclusiveness. The resulting phenomenological meaning of conscious technology and

[2] The positive complementarities signify the endogenous nature of the moral and material learning variables or 'de-learning' variables, as the case may be. The negative complementarities explain inter-variable marginal rate of substitution as in all of mainstream economics, borrowed by the so-called 'Islamic' economics. The selected variables thereby must be acceptable ones to explain the depth and purpose of moral-material inter-causal complementarities at the rejection or transformation of the inter-causal relations between non-inclusive variables. Choudhury qualifies such selective learning technologies as learning system of machines. See Choudhury, M.A. (2004). "Learning Systems", *Kybernetes: International Journal of Systems and Cybernetics, 33:1.*

technological change can then play the important role of conveying the integrated role of moral embeddedness into material goals for sustainable holism of the mind and materiality of economic gains. The theory of conscious technology can then be projected through such a moral-material unity of knowledge to attain wellbeing as the embodied goal.

The Holistic Model of Conscious Technology and Technological Change

In the context of our objective, the following methodological model is formalized. Pronouncedly, the methodology establishes the phenomenology of bringing together the moral and material construct of technology and technological change within the framework of participatory socio-economic development. The model of participatory development with moral-material embedding is the elemental groundwork of sustainability. It gains its methodological origin in the ontology and epistemology of unity of knowledge (Spencer, 2000). In Islamic methodological worldview, this ontological origin is the monotheistic explanation of the law of Tawhid and its relationship with the organic unity of being and becoming of the generality and specific of the world-system (Qur'an, 36:36). The complete phenomenological model of this Tawhidi (Islamic) methodology of monotheistic unity of knowledge is thereby coherent in the following parts:

The primal ontology of the Qur'an symbolized by the supercardinal mathematical domain[3] of Tawhid as law (Ω) and its mapping into the world-system via the medium of the sunnah (S). The formal ontological foundation is thereby denoted by, $\boldsymbol{\Omega} \rightarrow \rightarrow_{S} : (\boldsymbol{\Omega}, \mathbf{S})$.

The knowledge-flow of unity of knowledge in terms of the law emanate from the explication of (Ω, S) and as explained by the verses of the Qur'an. We denote this phase of derivation of unity of knowledge $\{\phi\}$ from $(\boldsymbol{\Omega}, \mathbf{S})$ by $(\boldsymbol{\Omega}, \mathbf{S}) \rightarrow \{\boldsymbol{\phi}\}$. $\{\phi\}$ includes both the fiqh ul-usul (jurisprudence) and fatawa (declarations), ra'y (opinions) and ihtisan (juristic preferences), pronounced by the mujtahids (shari'ah scholars) and tabiun (learned ones following the companions of the Prophet Muhammad) and the learned ones who in every age participate in deriving rules from (Ω, S) by way of discourse concerning the Tawhidi way of knowing 'everything' (a'lameen).

[3] Rucker, R. (1983). "Large cardinalities", in his *Infinity and the Mind*, Bantam Books, New York, NY.

Subsequently, understanding of organic unity of knowledge flowers out of $\{\phi\}$ to know how such search and discovery lead to the derivation of the worldly knowledge flows $\{\phi\}$ out of $\{\phi\}$ even as discourse and processes evolve out of continuous interaction, integration, and evolutionary learning (IIEC) (khalq in-jadid, Qur'an 10:34).[4] We denote such learning events by $\{\boldsymbol{\phi}\} \rightarrow \{\boldsymbol{\theta}\}$ as the worldly carrier of unity of knowledge in the grand design of the world-system.

$\{\boldsymbol{\theta}\}$ in turn simultaneously determines the unified world-system $\{\mathbf{x}(\boldsymbol{\theta})\}$ in the framework of its ontological principle of unity of knowledge that embodies all the knowledge-induced artefacts.[5] We write such a knowledge-induced vector in terms of its knowledge-induced variables in the {knowledge, knowledge-induced space, and knowledge-induced time} dimensions as, $\{\mathbf{x}(\theta)\} = \{x_1, x_2, \ldots, x_n\}[\theta]$, with n as a given limit of the indefinitely expanding number of divine blessings comprising the morally embedded materiality by $\{\boldsymbol{\theta}\}$, thereby, $\{\mathbf{x}(\boldsymbol{\theta})\}$.[6] We write, $\{\boldsymbol{\theta}\} \rightarrow \{\mathbf{x}(\boldsymbol{\theta})\}$. $\{\mathbf{x}(\boldsymbol{\theta})\}$ comprises all the endogenous variables that, in response to the Tawhidi ontology of unity of knowledge, also contains the policy and preference variables as endogenous ones as well. Time enters the vector as being knowledge-induced.[7] This means that, time in its absolute form $t(\theta)$ is simply a recorder, not a cause and determiner of knowledge-induced events. The events that now map the entire development processes are $\{E\{\theta, \mathbf{x}(\theta), t(\theta)\}\}$. The process-based continuity of such knowledge-induced events over time signifies the meaning of sustainability as the pervasively continuous moral-material inter-variable complementarities between the good things of life (halal al-tayyabah) at the divinely guided avoidance of the false things of life.

[4] Qur'an (10:34): "Say: 'Of your 'partners', can any originate creation and repeat it?" Say: "It is Allah Who originates creation and repeats it: then how are ye deluded away (from the truth)?"

[5] Qur'an (24:35): Allah is the Light of the heavens and the earth. The example of His light is like a niche within which is a lamp, the lamp is within glass, the glass as if it were a pearly [white] star lit from [the oil of] a blessed olive tree, neither of the east nor of the west, whose oil would almost glow even if untouched by fire. Light upon light. Allah guides to His light whom He wills. And Allah presents examples for the people, and Allah is Knowing of all things.

[6] Qur'an (55:67): "So which of the favors of your Lord would you deny?"

[7] Qur'an (65:12): It is Allah who has created seven heavens and of the earth, the like of them. [His] command descends among them so you may know that Allah is over all things competent and that Allah has encompassed all things in knowledge.

The moral-material vector now enters the ultimate goal of worldly application of wellbeing (Islamic maslaha W(E(θ)) in terms of Tawhidi unity of knowledge arising from the primal ontology of Tawhid as law. The maslaha function thereby measures the degree to which unity of knowledge exists in the estimated form of the system of circular causation relations between the selected variables for the problem under study. Or they are constructible by simulating the coefficients of the estimated equation system. The resulting evaluation of the wellbeing function subject to the circular causal endogenous relations, and thereby in respect of unity of knowledge and knowledge-induced variables, is formalized as follows:

Evaluate (estimation followed by simulation for unity of knowledge between variables):

Eval. $W(\theta) = W(E(\theta))$
Subject to circular causation relations,
$x_i(\theta) = f_i(\mathbf{x}_j,\theta,t)[\theta]$; $i, j = 1, 2, \ldots, n \leq N$; $i \neq j$.
The quantitative form of the wellbeing (maslaha) function is,
$\theta = F(\mathbf{x},\theta,t)[\theta]$

The combination of all the parts of the Tawhidi unity of knowledge-centred methodological worldview now yields the complete Process One of the continuous phenomenological model of moral-material endogenously embedded consciousness in the various sequential variables and relations. This model is applicable to the generality and particular of the evolutionary learning world-system. We write this complete phenomenological model as in expression (10.1) (Choudhury, 2017).

---------PROCESS ONE (IIEC)---------RECALL---------CONTINUE

$(\Omega,S) \to \{\phi\} \to \{\theta\} \to \{\theta,\mathbf{x}(\theta)\} \to \text{Eval.}W(\theta) = W(E(\theta))\ (\Omega,S)$ SUSTAINABILITY AND START NEW PROCESS (10.1)

Primal Ontology (TAWHID AS LAW)

Subject to circular causation relations,
$x_i(\theta) = f_i(\mathbf{x}_j,\theta,t)[\theta]$;
$i, j = 1, 2, \ldots, n \leq N; i \neq j$
$\theta = F(\mathbf{x},\theta,t)[\theta]$

Identifying Conscious Technology and Technological Change in Expression (10.1)

Technology and technological change can be represented by two such variables respectively in the vector $\{\mathbf{x}(\theta)\}$ that shows the embedding of moral values $\{\theta\}$ with the artefact of technology and technological change, $\{\mathbf{x}(.)\}$. The circular causation equations explain the moral-material combined role of technology and technological change in relation to the other variables that appear in the moral and material designations. For instance, technology and technological change are made to apply to goals like equity (x_e), fairness (x_f), and social justice (x_s) as moral values. They must also address poverty alleviation (x_p), life-fulfilment needs of development (x_d), business profitability (x_m), and national adaptation to indigenous expertise and values (x_a) as socio-economic variables. x_T and $dx_T/d\theta$ denote technology and technological change with respect to learning in terms of unity of knowledge 'θ', respectively.

The embedding moral and ethical values causing interrelationships between technology, technological change, and all the variables is the law of Tawhidi unity of knowledge. This parameter that is measurable in ranked quantitative values (Pratiwi et al., 2017) is denoted by 'θ'. Morality and ethics as synonymous values are explained by the parameters of complementarities while avoiding their opposite in negative complementarities, as of the neoclassical theory of marginal rate of substitution, and the social differentiation of social Darwinism.

The moral-material vector with all the component variables in natural logarithm is

$$\mathbf{x}(\theta) = \{x_e, x_f, x_s, x_p, x_d, x_m, x_a, x_T, dx_T / d\theta, \theta\}[\theta].$$

The technology-specific circular causation relation in natural logarithmic form to represent functional complementarities between the variables is,

$$\log(x_T) = f_T(\mathbf{x}(\theta), \theta);$$

$d\log T/d\theta = (1/T)(dT/d\theta)$ = rate of change in T with respect to'θ'. The sign of this variable is expected to be positive in the case of ethical meaning of technological change.

In the broader moral-material system of complementarities in the development vector $\mathbf{x}(\theta)$, positive coefficient signs of the technology-specific circular causation equation are expected, but are subject to simulation if needed to construct a desired moral-material relationship of the technology equation in respect of the other development variables. Such estimated and simulated coefficients represent partial elasticity of technology variable with respect to each of the development variables.

Sustainability

To answer the sustainability issue of continued recursive complementary interrelations between the endogenous variables in the whole system of inter-variable circular causation relations, it is expected, but are subject to simulation of the coefficients, that these coefficients will have positive signs. The technological effects in moral-material transformation of development with positive partial elasticity values in inter-variable circular causation relations conveys the meaning of technologically centred moral-material sustainability.

The incidence of sustainability thereby finds its defining presence in terms of the degrees of inter-variable complementarities in the wellbeing function (maslaha). The quantitative form of the wellbeing function was first conceptually expressed as $W(E(\theta) = \mathbf{x}(\theta))$, and is now quantitatively derived in the natural logarithmic form as,[8]

$$\theta = F\left(\mathbf{x}(\theta)\right)$$

[8] The derivation of the wellbeing function in the quantitative form as, $\theta = F(\mathbf{x}(\theta))$, from its conceptual form as $W(\theta) = W(\mathbf{x}(\theta))$ is done by considering that a linear approximation is affected to the general form, $W(\theta) = \Pi_{i=1}^{n} x_i(\theta)^{\theta}$. Thereby, using the Implicit Function Theorem of differential calculus with positive (non-zero) inter-variable specific elasticities (Baldani et al., 2005), we write the wellbeing function in the log-linear form, $\theta = F(\mathbf{x}(\theta))$. Furthermore, the theorem follows: A monotonic positive transformation of a utility (now wellbeing) function is a utility function (Henderson & Quandt, 1958). We abandon the use of the utility conception for the adoption of the wellbeing function with its altogether opposite implications based on pervasive inter-variable complementarities. The principle of pervasive inter-variable and inter-functional complementarities is derived from the organic dynamics of unity of knowledge arising from the primal ontology of Tawhid as the all-abiding law of generality and particularity in the world-system. See Barrow (1991) on unity of knowledge as consilience.

FINTECH IN THE WELLBEING MODEL OF UNITY OF KNOWLEDGE WITH TECHNOLOGY AND TECHNOLOGICAL CHANGE

FINTECH along with its recent technological entries, such as Bitcoin, cryptocurrency, online credit cards, and the like are simply mechanisms to optimize the flow of financial resources for customers and major shareholders in the financial world (Choudhury, 2018a). The sheer financial transaction and profit-optimizing objective of such mechanisms, despite the implied consequence in Islamic transactions in real asset-holding by means of Islamic financial instruments, do not make FINTECH an Islamic development item of the holistic goal of attaining sustainable Islamic technological change in Islamic development. On the contrary, the bare market-driven risky undertaking by the marginal clientele, who are usually risk-taking by their preferences to gain in market returns quickly, can cause uncertainties (risk-taking behavior). This kind of risk-taking preference can cause a bubble-bursting in the FINTECH industry.

FINTECH also rides the crest and trough of the business cycle where the urge to gamble the future is most intense, and thereby the risk-taking is deepest. On the other hand, smoothening of business cycle caused by declining unit risk in larger shareholding (stake holding) and effective risk and production diversification in the real economy may enhance the role of FINTECH in financial resource mobilization into real assets. This approach can be an actualization of Islamic mode of financing provided the moral-material embedding is observed. This test of FINTECH can be proved by empirically evaluating the FINTECH related wellbeing function subject to the system of circular causation relations.

What then ought to be the approach to risk and return in technological development in general that is exemplified by the introduction of financial engineering such as FINTECH, artificial intelligence, cryptocurrencies, and the like? It is noted that Islamic economics is presently devoid of any such institutional decision-making modulation in analytical terms. The risk-return approach in the light of the precept of wellbeing with the Islamic core principle of unity of knowledge is briefly formalized as follows:

Stakeholders' objective goal in asset-holding is to diversify risk while expecting stable and increasing returns from such holdings. Diversification (D) of risk is denoted by the formula,

$$D'(\text{risk diversification}) = R(\text{risk exposure}) / n(\text{number of stakeholders}).$$

Risk exposure is diversified further by spreading R over increasing production diversification in a joint production menu of development (q) that is diversified into q_i interrelated and thus compounded projects (sectors) as per the interactive, integrative, and evolutionary learning organic interrelations between them. Thereby, $q = \prod_{i=1}^{m} q_i^a$, where m denotes the number of diversified production outlets by projects (sectors), q_i. 'a' denotes the elasticity of 'q' to 'q_i'.

We then express D as a compound of risk and production diversification in respect of unit risk of asset holding,

$$D = (D'/n) * (1/q) = D' / \left[n * \prod_{i=1}^{m} q_i^a \right]$$

Furthermore, if n_i denote the specific number of stakeholders holding 'q_i' as unit project-specific returns, then we can write,

$$D = R / \left[\prod_{i=1}^{m} n_i . q_i^a \right]$$

The meaning is clear. It is not risk-aversion that guides the Islamic stakeholders' decision towards technological diversification D. Rather it is the unit risk of asset holding 'D' that encourages stakeholders to extend the willingness to expand investments. The return from such stakeholding of assets is denoted by $\prod_{i=1}^{m} n_i . q_i^a$. Now, as returns in the sense of production diversification by the joint production function of 'q' increases, D declines in the sense of unit risk-holding. The risk-return nature of Islamic investor and institutions would thus be characterized by risk-taking rather than risk-aversion, contrary to mainstream economic risk-return behavior. Besides, the nature of life-sustaining regime of development would call forth accessible stakeholders by offering them reasonably priced shares, unlike those of sukuk and FINTECH forms that assist mega-projects.

Monetary Resource Mobilization in the Space of Micro-money

A further institutional implication of micro-money such a cryptocurrencies for the above-mentioned regime of life-sustaining technological developments and risk and production diversification is the role of money as financial resource. To address this issue, we define micro-money resource mobilization by the project-wise equation of exchange of quantity theory of money as,

$$\Pi_{i=1}^{n} M_i v_i = \Pi_{i=1}^{n} Sp_i \Leftrightarrow M.v = Sp = p.y,$$

Sp_i denotes the ith project-specific spending as the function of money in circulation. M, v, and Sp are the aggregate variables formed by product-compounding, $\prod_{i=1}^{n}$. p denotes price level of transactions signified by the corresponding supply of output, y. Throughout the formal model $\prod_{i=1}^{n}$ means interaction, integration, and evolutionary learning organic interrelations between the corresponding variables as implied by the ontological precept of Tawhidi unity of knowledge by wellbeing. Wellbeing in turn is evaluated (estimated and simulated) by its endogenous circular causation relations between the attenuating variables and indicators. Note in the limiting case of full mobilization of monetary resources through the Islamic financing institutions and in the context of project specificity of monetary resource mobilization, $v_i \to 1$, $i = 1, 2, \ldots, m$.

In the case of an evolutionary learning money–real economy by the unity of knowledge denoted by induction of 'θ' - parameter of each of the endogenously related variables of wellbeing and its circular causation relations, any one of the following two conditions can occur:

$$dMv / d\theta = dSp / d\theta = 0 \Leftrightarrow d(p.y) / d\theta$$
$$= p(\theta).dy(\theta) / d\theta + y(\theta).dp(\theta) / d\theta = 0.$$

The meaning here is of a neutralized effect of technology and evolutionary learning denoted by 'θ' - endogeneity on monetary circulation in the money–real economy. Such is the case suspected of today's popularization of 'Islamic dinar' as a form of micro-money. Dinar as a monetary unit held by individual households and investors in large amounts remains ineffective in formulating an Islamic monetary regime and policy.

The other possibility gives,

$$dMv / d\theta = dSp / d\theta > 0 \Leftrightarrow d(p.y) / d\theta$$
$$= p(\theta).dy(\theta) / d\theta + y(\theta).dp(\theta) / d\theta > 0.$$

This form of the monetary effectiveness by technological and resource development is realized by the dynamic impact of 'θ' in all the variables as shown by further writing the above expression as,

$$dy(\theta) / y(\theta) > -dp(\theta) / p(\theta) \Rightarrow g(y(\theta)) > -g(p(\theta)).$$

The underlying meaning of this expression is that in a flourishing Islamic economy with micro-money in circulation, the resulting money–real economy finances a growing economic growth, $g(y(\theta))\uparrow$ that stabilizes decreasing inflationary tendency, $-g(p(\theta))$. In order to realize such continuous non-inflationary economic growth with the wellbeing implications of unity of knowledge by inter-variable organic causality of unity of knowledge, the dinar monetary system must be based on its monetary system not as private holding (Choudhury, 2018b). The effectiveness of dinar as monetary aggregate in Islamic monetary policy reform is inferred from the underlying organic unity of knowledge that endogenizes the money–real economy variables for sustainability of technological change in the economy-wide sense. This indeed is of a vast institutional significance of monetary transformation for sustainable technological change of the Islamic interconnected world-system.

Conclusion

This chapter has been a salient introduction on the theory of appropriate technology that endogenously combines the moral and material properties to serve the objective of wellbeing (Islamic maslaha). Such endogenous inter-variable relations are essentially proved by the complementary balance between the good goals that technology serves. One thereby notes that, in the vector of moral and material variables, all such variables are induced by the precept of unity of knowledge between the variables. This precept is derived uniquely from the primal ontology of Tawhid as the law of monotheistic organic unity of knowledge.

On the contrary, if we say that profit maximization as opposed to profits arising from the balance of evolutionary learning in circular causation relations feeds into the wellbeing function by statistical analysis, then the result would be maximization of the wellbeing function ceteris paribus, with all the other variables being given. Thereby, financial arrangements by the adoption of the specific technology in the perspective of technological change for socio-economic development would be treated as follows:

Max. $W(x_m, z)$, subject to given arrangement of interrelations between the variables as shown, yields,

$$dW = (\partial W / \partial x_m) dx_m + (\partial W / \partial z) dz = 0. x_T \textit{ is a variable within } \mathbf{z} \textit{ vector.}$$

Thereby we obtain, $(\partial W/\partial z)/(\partial W/\partial x_m) = -dx_m/dz$. This result implies the neoclassical rate of substitution between the technology variables and all the other variables. The presence of evolutionary learning is dissolved from technological choice. The role of ethics and the endogenous balance of inter-variable relations is annulled. In the choice of technology for socio-economic development with moral-material embedding of all the variables as complementary elements of evolutionary learning by unity of knowledge, any technology ought to maintain this arrangement to be an appropriate one. FINTECH would have to abide by the same principle to be acceptable in Islamic financial portfolio of mechanisms. The powerful role of transforming the monetary–real economy model of the Islamic world-system is formulated in terms of the monetary policy functioning of dinar. This kind of monetary–real economy transformation is of the institutional impact in the economy-wide sense, contrary to holding dinar privately as gold investiture.

References

Baldani, J., Bradfield, J., & Turner, R. W. (2005). *Mathematical Economics* (pp. 129–131). Canada: Thomson.

Barrow, J. D. (1991). Laws. In *Theories of Everything, the Quest for Ultimate Explanation* (pp. 12–30). Oxford: Oxford University Press.

Choudhury, M. A. (2004). Learning Systems. *Kybernetes: International Journal of Systems and Cybernetics, 33*, 1.

Choudhury, M. A. (2017). *A Phenomenological Theory of Islamic Economics*. Kuala Lumpur: University of Malaya Press.

Choudhury, M. A. (2018a). Micro-money, Finance and Real Economy Interrelationship in the Framework of Islamic Ontology of Unity of Knowledge and the World-System of Social Economy. *International Journal of Social Economics, 45*(2), 445–462.

Choudhury, M. A. (2018b). Comparative Islamic Perspectives in Money, Monetary Policy, and Social Wellbeing. *Journal of Economic Cooperation and Development, 39*(1), 143–162.

Henderson, J. M., & Quandt, R. E. (1958). *Microeconomic Theory, A Mathematical Approach*. New York, NY: McGraw-Hill Book.

Kaku, M. (2015). Consciousness—A Physicist's Viewpoint. In *The Future of the Mind*, Chapter 2. New York, NY: Anchor Book.

Pratiwi, A., Choudhury, M. A., & Ismal, R. (2017). *The Islamic Bank's New Paradigm Challenges and Blessings*. Germany: Noor Publishing.

Rucker, R. (1983). Large Cardinalities. In *Infinity and the Mind, the Science and Philosophy of the Infinite*. New York, NY: Bantam Books.

Spencer, N. (2000). *On the Significance of Distinguishing Ontology and Epistemology*. Contributed by Neville Spencer Upon Request, March 2000. See also *Critical Realism and Reality* by Andy Blunden, 2009. Internet Version.

CHAPTER 11

Endogenous Monetary Transmission in Islamic Financial Economics

Objective

This comprehensive picture is shown by the following sequence of effects: $S(r/i, y) = M(r/i, y) * v = p.\ y \Rightarrow 100\, per\, cent$ conversion of bank savings into money that is circulated fully via projects by financing instruments into the real economy. The symbols are defined as follows: S denotes bank savings as deposits that are held by banks as liquidity that is not mobilized into the real economy. 'r' denotes rate of return in the real sector mobilization of savings and money circulating via financing instruments. 'i' denotes rate of interest. 'r' and 'i' may be in nominal or real terms. 'y' denotes real output. M denotes real quantity of money as a function of the real values of y and r/i (Cagan, 1989). 'p' denotes the price level. 'v' denotes the velocity of circulation of money. As the circulation of a quantity of money in the above-mentioned relations equates with bank savings liquidated into real-economy financing (spending), then v tends to be 1 as the signal of full mobilization of bank savings into productive spending.

Professor Dr. Masudul Alam Choudhury and Dr. NirDukita Ratnawaty, Faculty of Economics, Trisakti University, Jakarta, Indonesia.

M. A. Choudhury, *Tawhid and Shari'ah*,
https://doi.org/10.1007/978-3-030-49087-4_11

Micro-Money Formalism in Islamic 100 Per Cent Reserve Requirement Monetary System

There are important social implications as endogenous ones linked with the economic ones of the above-mentioned relations. The fact of price stabilization with the perfectly elastic growth of the economy as a permanent phenomenon of the money, financing (spending), and real economy circular relations in respect of their endogenous inter-causal equations, means that the economic integration of this type has feedback with a life-fulfilment needs-economy and an increasingly privatizing economy. In the case of the macroeconomic general equilibrium system, this kind of adjustment is equivalent to an elastic form of non-inflationary economic growth and the complementary relations between aggregate demand with predominating spending (fiscal) stimulation and monetary policy induction (Friedman, 1953).

The project-specific expansion of monetary transmission complementing with the financing activity causes real output expansion. The results then are the market dominated transformation with public sector discursive coordination. Thereby, the real output expands along the intersection of the resulting aggregate demand and aggregate supply curves. Such a movement causes an elastic increase in productive employment and price-stability. In the Keynesian jargon (Mankiw, 2003) the full impact of the income multiplier is realized along such non-inflationary growth paths. Finally, price stability as a result also causes simultaneously a stability in (r/i) price relative.

Monetary Transmission Model of Islamic 100 Per Cent Reserve Requirement Monetary System

In Islamic real economy theory, the complementary coordination of money, finance, with the above-mentioned inner results in the variables causes reduction in 'i' even as the circulation of money, that is monetary transmission between banks and the real economy, enhances productive transformation. Figure 11.1 points out the nature of money and monetary transmission between the banks and the real economy in respect of liquidating the bank savings into productive mobilization by spending. The well-known theory of Money (M)-Commodity (C)-Money (M), thus MCM-model (Heilbroner, 1986) is now extended to $\{(MFRE)_{transmission\ 1} - (MFRE)_{transmission\ 2} etc\}(\theta)$.

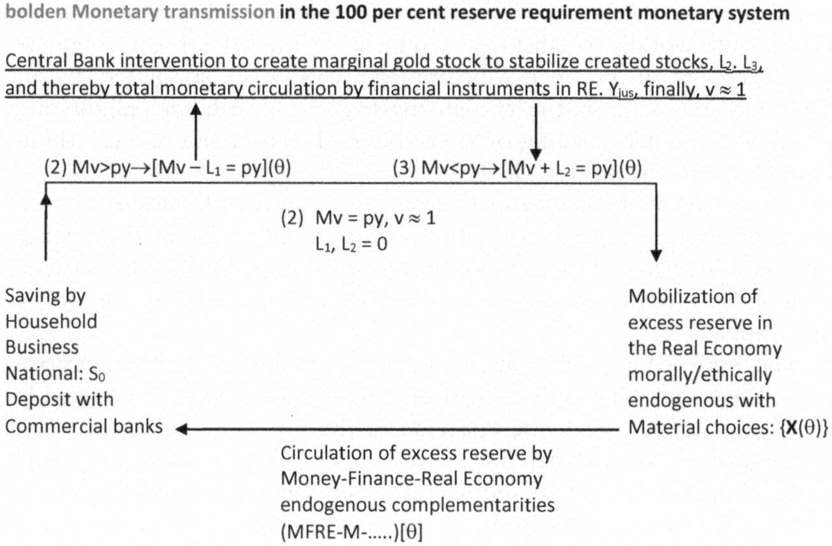

Fig. 11.1 Monetary transmission in 100 per cent reserve requirement monetary system of (MFRE) [**θ** endogenous complementary relations]

Symbolizing the Basic Model of Tawhidi Unity of Knowledge and Its Conceptualization in Money, Finance, and Real Economy Endogenous Interrelationship

Symbol 'θ' denotes consciousness parameter of knowledge based on the episteme of unity of knowledge. In Islamic epistemological worldview, the derivation of 'θ' is discursively determined by way of the foundational reference to the Islamic primal ontology of oneness of God (Tawhid). This premise is explained by the precept of unity of knowledge in endogenously circular causal interrelations between the good choices identified by their multidimensional variables. Briefly we write this foundational knowledge-deriving epistemology from the primal ontological roots as follows:

$\Omega \rightarrow_S \rightarrow \theta = plim$ {θ by discourse interactively with the learned ones} (Qur'an, 4:59).

Ω denotes the primal ontology as the Qur'an, emphasizing oneness of Allah represented as unity of knowledge by organic interrelations between the permissible choices while avoiding the impermissible ones (Qur'an, 36:36).

'S' denotes the transmission of the monotheistic knowledge in the Qur'an into worldly specifics denoted by 'θ'. 'S', called Sunnah, comprises the teachings of Prophet Muhammad, who Allah appointed by his Qur'anic command as the chosen messenger of Allah for enlightening access to the inner meaning of 'everything'. The meaning of 'everything' could be noted in Barrow (1991).

$\theta = \text{plim}\{\theta\}$ is the consensual derivation of unity of knowledge. It can be ordinally represented in the derived model of $(MFRE)[\theta]$ model by discourse in reference to the law of monotheism explained by the episteme of unity of knowledge as an organic process of knowing in the Qur'an and Sunnah.

The knowledge evolutionary discursive system of multidimensional evolutionary learning processes of diverse world-systems, all induced by knowledge arising from the episteme of unity of knowledge, and contrarily from the differentiated worldview of disjoint diversities not guided by $\Omega \rightarrow_S \rightarrow \theta$, both carry on inducing the generality and details of the specific world-systems under study. In this chapter, the specific world-system under study by the ontological and epistemological determination of $\Omega \rightarrow_S \rightarrow \theta$ is the $(MFRE - ....)[\theta]$ complementary problem under the guidance of 100 per cent reserve requirement monetary system.

The evaluation of the effectiveness of the organization and realization of the $(MFRE - - - -)[\theta]$, as explained above, is carried out by the wellbeing function, referred to in the methodology of $\Omega \rightarrow_S \rightarrow \theta = p \lim \{\theta\}$ as maslaha. The evaluation is made by estimation and simulation of this function, subject to the system of endogenously related inter-variable equations of circular causation relations in the light of the episteme of unity of knowledge, and vice versa as established in $\Omega \rightarrow_S \rightarrow \theta = p \lim \{\theta\}$. The maslaha function as wellbeing objective criterion is denoted by $W(MFRE, \theta)[\theta]$ with the inner variables.

Monetary Transmission in the 100 Per Cent Reserve Requirement Monetary System

Central bank intervention to create marginal gold stock to stabilize created stocks, L_2. L_3, and thereby total monetary circulation by financial instruments in RE. Y_{jus}, finally, $v \approx 1$.

A brief explanation of monetary transmission circulating in the 100 per cent reserve requirement monetary system is as follows. The initiation of savings is by productive activity in the national income, by way of which savings S_0 are deposited in the commercial banks. In the 100 per cent reserve requirement system, the savings become deposits in banks ready to be circulated as money via financial instruments (F) into the real economy, RE. Now there are three different scenarios in such a circulation of money by financial instruments into the real economy.

Firstly, the full amount of savings as deposits can circulate in the real economy by financing instruments aimed at specific projects. This kind of monetary circulation results in micro-money pursuing projects. Thereby, with the full mobilization of savings as deposits as money and financing there is the feedback consequence of v tending to unity. The contrary consequence is for the micro-money project-specific circulation to result in the case of fractional reserve requirement. This state of monetary relationship with the economy withholds the required deposits from full circulatory purpose of spending by the medium of micro-money. The underlying episteme of unity of knowledge is thus implied to prevail in sustaining this complementary or participatory organic relationship between money, finance (spending), and real economy $[MFRE](\theta)$.

This dynamic is represented by the parameterization of the unity of relations by θ - values (Choudhury, Pratiwi, Hossain, & Adenan in Toseef, forthcoming). The origin of such parameter-based inter-causal relations between the activities along with their inner variables is denoted by (MFRE)$[\theta]$; and M=py, with interrelated $M = A * \prod_{i=1}{}^{n} M_i^{ai}](\theta]$; $[p.\, y = spending = B * \prod_{i=1}{}^{n}(p.\, y)_i^{bi}][\theta]$. Both of these expressions apply in terms of project-specific micro-money and real-economy spending in them according to Islamic recommended choices (1 to n). Thus, the context of θ - *induction* appears.

The two cases of monetary transmission shown in Fig. 11.1 are, firstly, of excess quantity of money adjusted by a withdrawal of L_1 of quantity of money to restore the equation of exchange: $(Mv > py \rightarrow [Mv - L_1 = py]$ $(\theta))$. Secondly, the other case is of excess demand for money with adjustment. This requires injection of loanable fund by the central bank of L_1.

Thus, in every case, the adjusted equation of quantity theory of money and spending is restored. The result then is identical to the 100 per cent mobilization of money into the real economy via financial instruments and banking institutions. The central bank applies a marginal quantity of gold as numeraire to stabilize the currency value of all monies in circulation, including L_1, L_2, and the rest.[1]

The complementary dynamics of $[MFRE](\theta)$ prevails always. The result in monetary transmission then is 100 per cent reserve with commercial bank to match monetary flow through financial instruments with real economy activity. This is the meaning of 100 per cent reserve requirement monetary system. The central bank thereby ceases to have monetary governance on monetary transmission. Only marginally there exist the management of L_1 and L_2, and the overseeing and guidance of the states and policies of the economy with micro-money. In such money-finance-real economy interrelations the central bank establishes a discursive environment of consultation and policy guidance with commercial banks (Islamic banks), financial institutions, the real economy agencies, and all participants in economic activity.

A Focused Review of the Literature

100 per cent reserve requirement monetary transmission of the $[MFRE]$ (θ) type appears in the seminal works of Maritain (1985). Maritain's contribution on a nature of money recommends unlimited tokens representing money for circulation in the real economy, so that all individuals and households would benefit from these tokens by exchange of goods and services. Monetary control by the central bank, commercial banks, and other monetary authorities would disappear. In 100 per cent reserve requirement monetary system with the gold standard as a stable numeraire the circulation of micro-money is similarly recommended to be unlimited in quantity so as to be circularly related with the expanding real exchange in goods and services. Thereby, the wellbeing effect will be optimized and asset-backing by gold would be minimal in volume. Gold would be

[1] The dynamics of monetary transmission under 100 per cent reserve requirement monetary system imply that there is an inverse relationship between creation of gold (G) and trade (py). Thereby, gold creation comes to an end as the adjustments in limiting amounts of L_1 and L_2 restore the 100 per cent mobilization in $(MFRE)[\theta]$.

We write: $L_1 - L_2 = \Delta M = \Delta py = G$. We rewrite this expression as, $\Delta py/py = G/M$. That is, currency value of money in circulation, py, equals G/M. Now as monetary transmission increases, M increases in quantity, G declines in volume. Thereby, G/M = a, as a declining parameter. Change in currency value, g(py), is therefore stabilized around 'a' the total amount of money in circulation. That is G = a*M, with G declining asymptotically, as 'a' and M movement establish a rectangular hyperbola.

required by the central bank only to stabilize the circulatory currency value and the residual amount of excess reserves in the central bank (Fig. 11.1).

The praise for expansionary monetary supply for meeting the demand for spending in goods and services in the real economy was also proposed by the Austrian School of Economists (Yeager, 1997; Hayek, 1999). The base of micro-money was named as Unit. It was the numeraire to match up with a unit value of spending in goods and services in the real economy. On the side of the wellbeing contribution of such micro-money Schumpeter's (1961) entrepreneurial growth model is an example of micro-money being in high supply to finance sustainable microentrepreneurial projects.

In recent times, it is surmised by hind sight that the monetarism of an easy rather than a counter-cyclical contraction in the quantity of money into the real economy could have stemmed the tide of the Great Depression. In a growing economy, the complementarities between fiscal and monetary policies can be best fitted for non-inflationary economic growth.

In all such states of $[MFRE](\theta)$ model within the regime of 100 per cent reserve requirement monetary system, there remains the consistency of the effect of declining interest rates on the complementarities between money, finance, and real economy. Furthermore, the expanded release of monetary supply to endogenously support entrepreneurship, sustainability of non-inflationary economic growth, and thereby wellbeing. The early work of Friedman (1960) pointed out the possibility of such a holistic possibility of monetary arrangement.

The $[MFRE](\theta)$ model along with its early prototypes in the history of money, monetary policy, and institutional control over quantity of money for the real economy suggest that, this model is consistent with the state of a productive and ethically sustainable economic regime with wellbeing as monetary and financial resources increased to complement with the real economy. Thereby a circular causation exists between the dynamics of money, finance, and real economy interrelations and the phasing out of all forms of interest rates causing release of resources for sustaining the holistic consequences of the wellbeing criterion. This yields the situation whereby a dynamic evolutionary learning by endogenous interrelations between the inner variables of $[MFRE](\theta)$ type monetary transmission model is attained (Hayek, 1990).

Economic Stability by Means of the Relationship Between $[MFRE](\theta)$ and the Wellbeing Function

Economic stability caused by the $[MFRE](\theta)$ is explained in the case of the theory of monotheistic unity of knowledge in Islamic economics and finance by evaluation of the wellbeing function called the maslaha. Maslaha as wellbeing function parameterizes and explains the degree of pervasive positive complementarities that ought to exist between the variables denoting the good things of life and avoiding the negative complementarities with the respective variables signifying the undesired things. The maslaha objective function also has the properties of inter-variable complementarities of the two kinds. It is continuously differentiable because of the same nature of the learning parameter derived from the ontological and epistemological foundations of unity of knowledge of the monotheistic methodology. That is, $\theta = p \lim \{\theta\} \leftarrow (\Omega, S)$. Due to the properties of interactive, integrative, and evolutionary (IIE) learning of θ in the maslaha function (Choudhury, 2013), the maslaha function has non-optimal and non-steady state properties. Only evolutionary equilibriums caused by the force of 'θ' in generating estimated and simulated complementarities apply.

The ontological-epistemological (Spencer, 2000) origin of 'θ' - *induced* maslaha function also gives rise to endogenous circular causation relations between the maslaha variables. Thereby, the maslaha function is evaluated by estimation and simulation of coefficients in the 'as is' and 'as it ought to be' states of the problems under study in the light of pervasive complementarities between variables subject to the system of circular causation relations in evaluating the maslaha function.

Summarizing the Properties

One vector of selected variables in the $[MFRE](\theta)$ complementary problem in 100 per cent reserve requirement monetary system is $\mathbf{x}(\theta) = \{m, y, \mathbf{f}, p, e, r/I, \theta\}[\theta]$, with induction of all variables by 'θ'. The variables are, 'm' as real money. 'y' as real output, '**f**' as vector of financial instruments conducive of MFRE complementary relations in the circular flow of goods and finances, 'p' as output price signifying level stability, 'e' as employment, 'r/i' as relative between rate of return and interest rate, 'θ' as knowledge parameter inducing all these variables in terms of their inter-causal complementarities.

The conceptual maslaha function, subject to circular causation relations between the variables is $W(\theta, \mathbf{x}(\theta))$.

The circular causation relations between the variables are, $x_i(\theta) = A + \sum a_j(\theta) * \ln xj(\theta) + a(\theta) * \ln(\theta)$;
$i, j = 1, 2, \ldots, n; i \neq j$. Except for the last term, the θ - ***inductions*** of the variables and coefficients signify the sequential data of the variables corresponding to the θ - variables.

The quantitative form of the wellbeing function that can be evaluated (estimated in 'as is' state; and simulation in 'as it ought to be') is, $\theta = B + \sum a_i(\theta) * \ln x_i(\theta)$. This equation provides the sequence of $\mathbf{x}(\theta)$ - variables that are estimated (simulated) as predictor variables from the circular causation estimated (simulated) results in the variables.

The full description of the evaluation system of the maslaha function is given by,
Evaluate $W(\theta,\mathbf{x}(\theta)) = Const.\prod_{X_i}{}^{\alpha i*}\,\theta^{\beta}$.

This is converted to natural logarithmic form,

$$lnW(\theta,\mathbf{x}(\theta)) = Const.\sum \alpha_i * lnx_i + \beta * ln\theta \tag{11.1}$$

subject to the circular causation equations in natural logarithmic form,

$$lnx_i(\theta) = A + \sum a_j(\theta) * lnxj(\theta) + a(\theta) * \ln(\theta). \tag{11.2}$$

The quantitative form of the maslaha function to evaluate is,

$$\begin{gathered}\theta = B + \sum \mathrm{a_i}(\theta) * \mathrm{lnx_i}(\theta) \\ \mathrm{i, j} = 1, 2, \ldots, n; i \neq j\end{gathered} \tag{11.3}$$

Economic stability attained by the model of unity of knowledge generated by pervasive complementarities between the inner variables of $[MFRE](\theta)$ endogenous circular flow interrelations is explained further as follows. By some simplification of expression (11.1) in mathematical differentiation form we write in the form of rate of change, g(.) with respect to 'θ' in view of degrees of inter-variable complementarities as,

$$g(W) = \sum_i \alpha_i * g(x_i) + \beta * g(\theta) > 0 \tag{11.4}$$

Now because every variable is positively complementary with θ - value, therefore, economic stability in the midst of ethically induced variables of the $[MFRE](\theta)$ model is implied.

The explanation of economic stability is made by the ontology, epistemology, and phenomenology of Tawhid as law in respect of the Qur'anic meaning of monotheism and its methodological representation in terms of unity of knowledge between recommended complementary variables. Yet this result is not true of the shari'ah-based framework of Islamic economics and finance. Two examples are given here to establish the inability of shari'ah, fiqh, and fatwa related misguided claim of economic stability.

Firstly, we take the case of the absence of ijara-type pricing, such as rental in ijara (assets acquired by deferred payments), premiums (as of takaful), deferred payments (as of mark-up pricing by murabaha), and risk and return pricing on credit sale (as in the case of twarruq, innah, and istisna). There are also the many cases that are daily conjured up by the so-called 'shari'ah-compliance' legitimation of Islamic financial instruments.

All such deferred rental kinds of rates as prices on financial instruments, and thereby their pricing relations with goods and services in market exchange, are indeterminate. This problem is true at the initial point of a mutual contract, unless such prices are forward-looking prices that can be realized only 'nearest' to the point of exchange transactions. Forward-looking pricing is tantamount to interest rate (riba), for it remains undetermined at the present time of a future contract. Thereby, real economic transactions remain equally undetermined in valuation. On the other hand, if goodwill of donations is used to form financial capital, this is not feasible under continued conditions of uncertainty and financial limitations of groups and individuals as donors for free-ridership. Feasibility and sustainability demand organized forms of pricing for risk, return, and transactions in the intertemporal sense. No formula has been derived according to shari'ah to resolve this critical outstanding problem of forward-looking ijara type pricing. There is no evidence of shari'ah instruments regarding their stability role in financing during global financial crises, as otherwise claimed. The only other way how this could happen is by relying upon large inflow of rich shareholders' funds. Such a case was implied in the greater than unit value of Finance/Deposit ratio in shareholding of Islamic banks during the financial crises.

Twarruq as an Example of Micro-Money in Islamic Financial Mobilization

One of the kinds of micro-money that has been debated by shari'ah-compliance institutional financial innovations in recent times is called twarruq. Twarruq is a short-selling financial credit arrangement. The buyer of an asset buys on credit without paying right then. Upon financial need, the buyer sells the credited asset at a lower price to raise the needed funds. In the light of the Tawhidi law of unity of knowledge derived from the monotheistic origin of Islamic worldview, twarruq is both ridiculous and unlawful and inefficient mode of financing. Thereby, the use of such a fund is contrary to achieving economic stability. How and Why? The principle of relational unity between Money (M), Finance (F), Real Economy (RE), and thereby such sustained continuity by the $[MFRE - M....](\theta)$ symbiotic relationship is foregone. The ijara-type pricing problem of intertemporal interest rate, as mentioned above, reappears and remains unresolved.

It can be explained that twarruq causes devaluation of assets by lowering the effective prices in short-selling of assets. There also appears a smaller available financial need to meet the increasing asset transaction demand. The economy thereby comes to have two adverse effects. Lesser amounts of funds circulated into the economy by asset devaluation cause a less productive economy with lower funds flow. The explanation of such marginal substitution between credited assets and their short-selling at lower price follows a neoclassical approach.[2]

All the tenets of complementarity of $[MFRE](\theta)$ are thus lost. The Tawhidi law of relational unity superseding the legitimacy given to the differentiated ways of shari'ah is violated. The shari'ah, fiqh, and fatawa (jurisprudence, opinionated religious verdicts) in all such various cases of pricing and asset valuation and dynamic resource allocation by inter-causal learning between the critical variables are untenable.[3]

[2] A 'marginal substitution' form of the neoclassical production possibility curve between credited assets (superior, A) and cheaply sold same asset to raise cash (inferior, A') can be shown along the shifting production possibility curves with resource allocation increasingly into A'. The respective prices being p and p'. We note the marginalist formula of 'optimal' resource allocation along the shifting production possibility curve caused by optimal resource allocation along the inferior asset curve, A'-trajectory: $-(\Delta A/\Delta A') \downarrow = (p'/p)\downarrow$, as $\Delta A\downarrow$ and $\Delta A'\uparrow$. Thereby, $(p'/p)\downarrow$. Twarruq allocation of resources abides in this case of marginalism along the inferior goods trajectory for the credited item by the short-selling items. The (MFRE) circular relations in continuity are annulled. The truly Islamic epistemological inference of unity of knowledge between these variables is annulled in the cantankerous shari'ah approach. All the implications of social and economic stability arising from the wellbeing objective criterion are annulled.

[3] On the contrary, the neoclassical result is negated in the precept of unity of knowledge with interactive, integrative, and evolutionary learning dynamics. Now 'θ' as knowledge-inducing parameter of complementarities between A and A' causes both of these to increase continuously along with the increase in resource allocation by learning. Thereby, $(p'/p)[\boldsymbol{\theta}] > 0$, also causes $(\Delta A/\Delta A')[\boldsymbol{\theta}] > 0$, along the continuously increasing resource allocation with θ - ***induction*** increasing in the neighbourhood of probabilistic occurrence of comple-

Empirical Results of Islamic Monetary Transmission Model for Bank Indonesia

A monetary transition model for the Bank Indonesia, defined as the degree of effectiveness of mobilizing depositors' funds as bank savings into productive transformation of the real economy, has been estimated and statistical results derived. The real economy is further disaggregated for the primary sector, secondary sector, and tertiary sector. The variants of such a model are given below (NirDukita, 2018). In terms of the $[MFRE](\theta)$, the Tawhidi methodological tenets of continuity in respect of the Tawhidi episteme of unity of knowledge is satisfied if there are sustained complementarities in the suggested model between their inner variables. The vector of variables would comprise socio-economic variables, financial instruments of resource mobilization, and epistemic value parameters denoted by 'θ'. The specific methods to evaluate the resulting statistical models of $[MFRE](\theta)$ with their inner endogenously interactive, integrative, and evolutionary (IIE) learning variables, as adduced in the wellbeing function of inter-variable complementarities signifying Tawhidi episteme of monotheistic unity of knowledge, are tried out. The endogenous inter-variable equations form the circular causation models signifying degrees of interactive integration attained by the force of evolutionary learning. Finally, the wellbeing function for $[MFRE](\theta)$ for the three real economy sectors are determined. Thus, all the phenomenological attributes of Tawhidi methodological worldview are activated for a critical examination of the $[MFRE](\theta)$ model. This is followed by its recommendations in the monetary transmission model tested for Bank Indonesia.

Description of variables (symbols) in the monetary transition model:

1. PS (primary sector) denotes the total of sectoral GDPs for the following sectors—agriculture, mining, processing industry, electricity, gas, and construction.
2. Y (real GDP) is based on constant prices of 2010.
3. PS/Y denotes the ratio of primary sector GDP to real GDP (real output).

mentary inter-variables relations of (MFRE)[$\boldsymbol{\theta}$]. The neoclassical results of optimal resource allocation and marginal rate of substitution (opportunity cost) are annulled in the Tawhidi dynamics of unity of knowledge along all continuums of complementary allocation of resources between the selected variables. Twarruq-type allocation is dispelled from acceptance in the Tawhidi methodological worldview by the epistemological dynamics of $\boldsymbol{\theta}$ - ***induction*** of complementarities in the (MFRE)[$\boldsymbol{\theta}$] circular causation relations of the wellbeing objective criterion. Thereby, economic and social stability is attained, as explained in the text.

4. SS (secondary sector) denotes the total of GDPs of the trade, transportation, and business services obtained by summing up the GDPs for the trade, hotel, and restaurant; transport and communication; finance, real estate, and business services sectors.
5. SS/Y denotes the ratio of secondary sector GDP to real GDP (real output).
6. TS (tertiary sector) denotes the total of GDPs for services (government and private services).
7. TS/Y denotes the ratio of tertiary sector GDP to real GDP (real output).
8. Fin/M denotes the ratio of total financing by Shari'ah banks in Indonesia to total money supply (M0+M1).
9. SIMA denotes Interbank Mudarabah Investment.
10. SIMA/TotFinde notes the ratio of SIMA to total financing by shari'ah banking in Indonesia interpreted as portion of liquid funds in money market.
11. Theta 'θ' denotes the knowledge variables. 'θ' is obtained by determining the parametric weight of data for the variables. The weight of each of the different variables is based on an understanding of the highest rank, say 10, assigned for the best data values in the light of Tawhidi methodology (Choudhury & Rahim, 2016). The formula for calculating values in each column is:
    θ = (10/*corresponding observation value.* θ = 10 for best value of the variable) * (observed variable values in the column). After getting all computations of such weights from each column variables, they are averaged across columns by rows to obtain the average 'θ' of all variables.

Results of Data Analysis

Simultaneous Ordinary Least Square (OLS)

Hereinafter, the prerequisite test results presented from the entire regression model are as follows:

Normality Test

The results obtained from the test of the normality assumption with the Jarque-Bera normality test for transmission model data show that the error follows the normal distribution with ρ - *value* values of 0.378, which is greater than 0.01. This means that with a 99 per cent confidence level it can be said that the error term is normally distributed.

Multicollinearity Test
Results of tests on the assumption of absence of multicollinearity showed that in the monetary transmission mechanism model no presence of multicollinearity is noted. This is indicated by the correlation coefficient between each independent variable in each model being smaller than 0.8.

Heteroscedasticity Test
Homoscedasticity test results with white heteroscedasticity in the transmission mechanism model showed ρ value square of 0.2863, that is, greater than 0.01. It can therefore be concluded that no heteroscedasticity exists in the regression model with a 99 per cent confidence level.

Autocorrelation Test
Here the results are obtained from the tests on the assumption of the absence of autocorrelation with the Breusch-Godfrey. Serial Correlation LM Test showed that the regression model contained autocorrelation. The value of ρ - *value* in the data of transmission model is null. This value being less than 0.01, means that there is autocorrelation in the regression model.

The monetary transmission model is further estimated by the VAR method to accommodate lag effect, especially by the use of VAR-Granger Causality to get the next picture from the transmission mechanism path.

Vector Autoregressive (VAR)

The VAR stability test is required to determine the stability of VAR estimation. The results of the VAR stability test in the following models show that a stable VAR model is indicated by the modulus value being less than one (Gujarati, 2010).

Discussion of the Monetary Transmission Model of [MFRE](θ)

The [MFRE](θ) monetary transmission model uses the Islamic methodology approach with a view of the unity of divine knowledge (Choudhury, 2010).This is the model that is always anchored in the Tawhidi methodology based on the Quran and sunnah. The Qur'an was revealed to the

Prophet to bring grace and welfare to all the universes (Qur'an, 21:107). The Qur'an presents a model of development within the framework of wellbeing system with the spirit of participation, paired in a process of circular causation (Qur'an: 42:38). The strong base is circular causation in the evaluation of wellbeing (maslaha) according to the Tawhidi episteme of unity of knowledge. The relations of circular causation result in sustainability explained by simulations along the Tawhidi String Relation (TSR) up to the Hereafter. In a saying of Prophet Muhammad, it is declared that the search and discovery of the sources of the Qur'an and sunnah will prevail until the end of time. The Qur'an provides a conceptual basis of analysis involving the inherent morality in economics, society, and government (Qur'an 2: 261–83). In its implementation, the complementary participatory concept between the variables of each circular causation equation is accommodated by including the parameter θ in each of the included variables. The θ value of each estimated relation is the average of the weighted index θ inducing each variable in the [MFRE](θ) monetary transmission model.

Furthermore, the discussion of the results according to this Islamic framework follows the stages of analysis given here:

1. Using Ordinary Least Square Analysis Methods for Effectiveness Analysis, the estimation result of monetary transmission mechanism model refers to the mainstream econometric concept, which does not involve 'θ' as an indicator of the degree of complementarities, which symbolizes learning between the variables in the model. The [MFRE](θ) monetary transmission mechanism model includes θ to represent the presence of the learning process of each economic agent (represented by each variable) in the model. The first evaluation method used is the simultaneous OLS method, firstly without and next by entering the 'θ' - effect to give comparison between these two aspects of the monetary transmission model, firstly without and then with learning process.
2. Next the Vector Autoregressive (VAR) with Granger Causality method is used to get a transmission mechanism path. This model is a dynamic model of quantitative behavioral type.

Empirical Analysis of the Monetary Transmission Model without 'θ' - parameter

The estimation result of the monetary sector linkage model to the real economy by three sectors is carried out without the learning parameter 'θ'. Statistical software Eviews 9 yields the following regression estimation in natural loglinear form (LN):

Primary Sector

$$LNPS\_Y = 0.99444 - 0.08993 * LNFIN\_M - 0.00216 * LNSIMA\_FIN - 0.00017 * RSBIS + 1.43272e-05 * INF$$

$$t\text{-}stat: (632.8512)***(-28.1313)***(-2.9518)**(-6.4277)*** (0.1528)$$

$$R^2 = 0.924073; DW = 0.284522 \quad (11.5)$$

$$LNFIN\_M = 9.66382 - 9.65612 * LNPS\_Y - 0.03158 * LNSIMA\_FIN - 0.00129 * RSBIS - 0.00037 * INF$$

$$t\text{-}stat: (29.5546)***(-28.1313)***(-4.3202)**(-4.2606)*** (-0.3791)***$$

$$R^2 = 0.916643; DW = 0.263233 \quad (11.6)$$

$$LNSIMA\_FIN = 32.38641 - 31.32116 * LNPS\_Y - 4.26216 * LNFIN\_M - 0.01621 * RSBIS - 0.02004 * INF$$

$$t\text{-}stat: (3.0778)***(-2.9518)***(-4.3202)**(-4.6535)*** (-1.7992)*$$

$$R^2 = 0.390041; DW = 0.454567 \quad (11.7)$$

$$RSBIS = 1455.45599 - 1469.08676 * LNPS\_Y - 101.54637 * LNFIN\_M - 9.42822 * LNSIMA\_FIN - 0.18678 * INF$$

$$t\text{-}stat: (6.3942)***(-6.4277)***(-4.2606)**(-4.6535)***(-0.6877)$$

$$R^2 = 0.510917; DW = 0.335029 \quad (11.8)$$

(continued)

(continued)

$$INF = -10.17250 + 13.58620 * LNPS\_Y - 3.250919 * LNFIN\_M$$
$$-1.31103 * LNSIMA\_FIN - 0.02102 * RSBIS$$
$$t\text{-}stat: (0.1151)(0.1528)(-0.3791)(-1.7992)*(-0.6877)$$
$$R^2 = 0.041533; DW = 1.183478 \quad (11.9)$$

Note: ***denotes significance at 99 per cent confidence level

** denotes significance at 95 per cent confidence level

*denotes significance at 90 per cent confidence level

No asterisk (*) denotes not significant

The coefficients of each equation show the magnitude of elasticity in relation to the endogenous dependent variable in circular causation. The above equations can be rewritten in the Cobb-Douglas form:

$$PS\_Y = 4{,}199\ Fin\_M^{-0{,}0899} Sima\_Fin^{-0.0022} rSBIS^{-0{,}0002} Inf^{0.00001} \quad (11.10)$$

$$Fin\_M = 15737{,}47 PS\_Y^{-9{,}6561} Sima\_Fin^{-0.03158} rSBIS^{-0{,}0013} Inf^{0.0004} \quad (11.11)$$

$$Sima\_Fin = 1{,}16E + 14\ PS\_Y^{-31{,}3212} Fin\_M^{-4{,}2621} rSBIS^{-0{,}02} Inf^{0.02} \quad (11.12)$$

$$rSBIS = PS\_Y^{-1469{,}08} Fin\_M^{-101{,}54} Sima\_Fin^{-9{,}428} Inf^{-0.1867} \quad (11.13)$$

$$INF = 3{,}38E - 05\ PS\_Y^{13{,}58} Fin\_M^{-3{,}251} Sima\_Fin^{-1{,}311} rSBIS^{-0{,}02} \quad (11.14)$$

Secondary Sector

$$LNSS\_Y = 0.81899 + 0.23129 * LNFIN\_M + 0.00236 * LNSIMA\_FIN$$
$$+0.000366 * RSBIS - 7.815456e - 06 * INFt - stat: (276.3376)***(38.3602)***$$
$$(1.7053)*(7.0387)***(-0.0442)$$
$$R^2 = 0.957741;\ DW = 0.247737 \quad (11.15)$$

(continued)

(continued)

$$LNFIN\_M = -3.23895 + 3.99779 * LNSS\_Y - 0.01555 * LNSIMA\_FIN$$
$$-0.00122 * RSBIS - 0.00026 * INF$$
$$t\text{-}stat: (-33.5036)***(38.3602)***(-2.7595)***(-5.3563)***$$
$$(-0.3522)$$
$$R^2 = 0.952266; DW = 0.238512 \quad (11.16)$$

$$LNSIMA\_FIN = -6.93076 + 10.04756 * LNSS\_Y - 3.8376 * LNFIN\_M$$
$$-0.01488 * RSBIS - 0.02137 * INF$$
$$t\text{-}stat: (-1.4301)\ (1.7053)***(-2.7595)***(-3.9869)***$$
$$(-1.8779)*$$
$$R^2 = 0.361234; DW = 0.435553 \quad (11.17)$$

$$RSBIS = -669.8449 + 811.5558 * LNSS\_Y - 158.615988 * LNFIN\_M$$
$$-7.86099 * LNSIMA\_FIN - 0.1914 * INF$$
$$t\text{-}stat: (-7.1088)***(7.0387)***\ (-5.3563)***(-3.9869)***$$
$$(-0.7227)$$
$$R^2 = 0.534652; DW = 0.347672 \quad (11.18)$$

$$INF = 5.04529 - 2.08377 * LNSS\_Y - 3.9915 * LNFIN\_M$$
$$-1.3357 * LNSIMA\_FIN - 0.02264 * RSBIS$$
$$t\text{-}stat: (0.1306)\ (-0.0442)(-0.3522)(-1.8779)*(-0.7227)$$
$$R^2 = 0.41362; DW = 0.183129 \quad (11.19)$$

The Cobb-Douglass form is written as follows:

$$SS\_Y = 2{,}265\ Fin\_M^{0.231} Sima\_Fin^{0.002} rSBIS^{0{,}0003} Inf^{-0{,}.000007} \quad (11.20)$$

$$Fin\_M = 0{,}039\ SS\_Y^{3{,}997} Sima\_Fin^{-0.015} rSBIS^{-0{,}0012} Inf^{0.0003} \quad (11.21)$$

$$Sima\_Fin = 0{,}0009\, SS\_Y^{10{,}047} Fin\_M^{-3{,}837} rSBIS^{-0{,}0148} Inf^{0.021} \quad (11.22)$$

$$rSBIS = 1,2E - 291SS\_Y^{881,55} Fin\_M^{-158,615} Sima\_Fin^{-7,860} Inf^{-0.191} \tag{11.23}$$

$$INF = 155,244\, SS\_Y^{-2,083} Fin\_M^{-3,991} Sima\_Fin^{-1,335} rSBIS^{-0,0226} \tag{11.24}$$

Tertiary Sector

$$\begin{aligned} &LNTS\_Y = 0.7825 + 0.0818 * LNFIN\_M + 0.0086 * LNSIMA\_FIN \\ &-1.1412e-05 * RSBIS + 0.0005 * INF \\ &t\text{-}stat: (59.6468) * * * * (3.0629) * * * (1.4107)(-0.0504) \\ &(-0.5797)\, R^2 = 0.081240; DW = 0.136003 \end{aligned} \tag{11.25}$$

$$\begin{aligned} &LNFIN\_M = -0.2614 + 0.88700 * LNTS\_Y - 0.0831 * LNSIMA\_FIN \\ &+0.0028 * RSBIS - 0.0040 * INF \\ &t\text{-}stat: (-1.0982)(3.0629) * * * (-4.4124) * * * \\ &(3.9203) * * * \quad (-1.5508)\, R^2 = 0.412828; DW = 0.085567 \end{aligned} \tag{11.26}$$

$$\begin{aligned} &LNSIMA\_FIN = -0.1725 + 1.8917 * LNTS\_Y - 1.6799 * LNFIN\_M \\ &-0.0113 * RSBIS - 0.0225 * INF \\ &t\text{-}stat: (0.1604)\ (1.4107)(-4.4124) * * * (-3.5495) * * * \\ &(-1.9659) * R^2 = 0.356427; DW = 0.398679 \end{aligned} \tag{11.27}$$

$$\begin{aligned} &RSBIS = -5.8859 - 1.8544 * LNTS\_Y + 41.2349 * LNFIN\_M \\ &-8.3899 * LNSIMA\_FIN - 0.2786 * INF \\ &t\text{-}stat: (-0.2011)\ (-0.0504)(3.9203) * * * (-3.5495) * * * (-0.8846) \\ &R^2 = 0.342542; DW = 0.219064 \end{aligned} \tag{11.28}$$

$$\begin{aligned} &INF = -1.4131 + 6.0608 * LNTS\_Y - 4.9569 * LNFIN\_M \\ &-1.3893 * LNSIMA\_FIN - 0.0233 * RSBIS \\ &t\text{-}stat: (-0.1670)\ (0.5797)(-1.5508)(-1.9659) * (-0.8846) \\ &R^2 = 0.043942; DW = 1.183592 \end{aligned} \tag{11.29}$$

The Cobb-Douglass form is:

$$TS\_Y = 2{,}186 Fin\_M^{0{,}082} Sima\_Fin^{0.008} rSBIS^{-1{,}141e-05} Inf^{0{,}.0005} \tag{11.30}$$

$$Sima\_Fin = 0{,}841 TS\_Y^{1{,}891} Fin\_M^{-1{,}679} rSBIS^{-0{,}011} Inf^{-0023} \tag{11.31}$$

$$rSBIS = 0{,}0027 TS\_Y^{-1{,}854} Fin\_M^{41{,}23} Sima\_Fin^{-8{,}389} Inf^{-0{,}278} \tag{11.32}$$

$$INF = 0{,}2434 TS\_Y^{6{,}06} Fin\_M^{-4{,}9561} Sima\_Fin^{-1{,}389} rSBIS^{-0{,}023} \tag{11.33}$$

Monetary Transmission Model as Linkage Model of the Monetary Sector with the Three-Sector Real Economy with θ - Parameterization

The estimation results of the monetary sector linkage model to the sectoral domestic product processed using Eviews 9 using 'θ' - parameterization yields the following regression equations:

Primary Sector

$$LNPS\_Y = 0.9615 - 0.0474 * LNFIN\_M - 0.0062 * LNSIMA\_FIN$$
$$+0.0004 * RSBIS + 0.0008 * INF + 0.0033 * \theta 1$$
$$t\text{-}stat: (332.4868)***(-11.6095)***(-10.4715)***(8.1853)***$$
$$(8.6713)***(12.2111)***$$
$$R^2 = 0.953301; DW = 0.403580 \tag{11.34}$$

$$LNFIN\_M = 11.0913 - 11.1985 * LNPS\_Y - 0.04747 * LNSIMA\_FIN$$
$$+0.00037 * RSBIS + 0.0020 * INF + 0.0105 * \theta 1$$
$$t\text{-}stat: (12.3785)***(-11.6095)***(-4.0258)***(0.3614)$$
$$(1.1947)(1.7090)*$$
$$R^2 = 0.918640; DW = 0.275288 \tag{11.35}$$

(*continued*)

(continued)

$$LNSIMA\_FIN = 73.7758 - 77.6088 * LNPS\_Y - 2.5250 * LNFIN\_M$$
$$+0.0540 * RSBIS + 0.0809 * INF + 0.3894 * \theta 1$$
$$t\text{-}stat: (10.2359)***(-10.4715)***(-4.0258)***(9.7632)***$$
$$(8.02519)***(13.7971)***$$
$$R^2 = 0.765370; DW = 0.418890 \quad (11.36)$$

$$RSBIS = -768.1288 + 830.0516 * LNPS\_Y + 2.9799 * LNFIN\_M$$
$$+8.2299 * LNSIMA\_FIN - 1.3407 * INF - 5.8088 * \theta 1$$
$$t\text{-}stat: (-7.7128)***(8.1853)***(0.3614)(9.7632)***$$
$$(-14.3349)***(-32.5914)***$$
$$R^2 = 0.950723; DW = 0.391352 \quad (11.37)$$

$$INF = -476.9551 + 510.8124 * LNPS\_Y + 5.8161 * LNFIN\_M$$
$$+4.3406 * LNSIMA\_FIN - 0.47235 * RSBIS - 3.0083 * \theta 1$$
$$t\text{-}stat: (-8.2657)***(8.6713)***(1.1947)(8.02519)***$$
$$(-14.3349)***(-16.0713)***$$
$$R^2 = 0.697690; DW = 0.788236 \quad (11.38)$$

***Note: denotes significance at 99 per cent confidence level

**denotes significance at 95 per cent confidence level

*denotes significance at 90 per cent confidence level

No asterisk (*): Not significant

The Cobb-Douglass form is:

$$PS\_Y = 2{,}6156 Fin\_M^{-0{,}047} Sima\_Fin^{-0{,}006} rSBIS^{0{,}0004} Inf^{0.0008} \theta 1^{0.003} \quad (11.39)$$

$$Fin\_M = 65597{,}96 PS\_Y^{-11{,}199} Sima\_Fin^{-0.047} rSBIS^{-0{,}0003} Inf^{0.002} \theta^{0.01} \quad (11.40)$$

$$Sima\_Fin = 1{,}09E+32\, PS\_Y^{-77{,}608} Fin\_M^{-2{,}525} rSBIS^{-0{,}054} Inf^{0.08} \theta^{0.389} \quad (11.41)$$

$$rSBIS = PS\_Y^{380,052} Fin\_M^{2,979} Sima\_Fin^{8,229} Inf^{-1.341} \theta^{-5,808} \quad (11.42)$$

$$INF = 7,29E - 208\, PS\_Y^{510,81} Fin\_M^{5,82} Sima\_Fin^{4,34} rSBIS^{-0,472} \theta^{-3,008} \quad (11.43)$$

Secondary Sector

$$\begin{aligned} &LNSS\_Y = 0.8367 + 0.13657 * LNFIN\_M - 0.0075 * LNSIMA\_FIN \\ &+0.00112 * RSBIS + 0.0013 * INF + 0.0057 * \theta 1 \\ &t\text{-}stat: (231.7306)***(9.2824)***(-4.0491)***(9.4165)*** \\ &(5.3583)***(6.8668)*** \\ &R^2 = 0.969734; DW = 0.238579 \end{aligned} \quad (11.44)$$

$$\begin{aligned} &LNFIN\_M = -2.4447 + 3.0751 * LNSS\_Y - 0.0357 * LNSIMA\_FIN \\ &+0.0009 * RSBIS + 0.0028 * INF + 0.0132 * \theta 1 \\ &t\text{-}stat: (-8.5087)***(9.2824)***(-4.0590)***(1.1541) \\ &(2.2031)**(2.92445)*** \\ &R^2 = 0.955467; DW = 0.283460 \end{aligned} \quad (11.45)$$

$$\begin{aligned} &LNSIMA\_FIN = 14.9043 - 16.1188 * LNSS\_Y - 3.4042 * LNFIN\_M \\ &+0.0547 * RSBIS + 0.0831 * INF + 0.3988 * \theta 1 \\ &t\text{-}stat: (4.5383)***(-4.0491)***(-4.0590)***(10.3436)*** \\ &(8.3636)***(14.5386)*** \\ &R^2 = 0.769916; DW = 0.447484 \end{aligned} \quad (11.46)$$

$$\begin{aligned} &RSBIS = -330.7727 + 380.7830 * LNSS\_Y + 12.9198 * LNFIN\_M \\ &+8.656 * LNSIMA\_FIN - 1.369 * INF - 5.9243 * \theta 1 \\ &t\text{-}stat: (-10.0512)***(9.4166)***(1.1541)(10.3436)*** \\ &(-14.3983)***(-31.3656)*** \\ &R^2 = 0.949786; DW = 0.395021 \end{aligned} \quad (11.47)$$

(continued)

(continued)

$$INF = -130.8673 + 149.5695 * LNSS\_Y + 14.153 * LNFIN\_M$$
$$+4.456 * LNSIMA\_FIN - 0.4640 * RSBIS - 3.0351 * \theta 1$$
$$t\text{-}stat: (-5.6593)***(5.3583)***(2.2031)**(8.3636)***$$
$$(-14.3983)***(-16.306)***$$
$$R^2 = 0.703628; DW = 0.812723 \quad (11.48)$$

The Cobb-Douglass form is,

$$SS\_Y = 2{,}308 Fin\_M^{0{,}136} Sima\_Fin^{-0{,}007} rSBIS^{0{,}0011} Inf^{0.0013} \theta 1^{0.005} \quad (11.49)$$

$$Fin\_M = 0{,}086 SS\_Y^{3.075} Sima\_Fin^{-0.0357} rSBIS^{-0{,}0009} Inf^{0.0028} \theta^{0.013} \quad (11.50)$$

$$Sima\_Fin = 2970676\, SS\_Y^{-16{,}11} Fin\_M^{-3{,}404} rSBIS^{0{,}054} Inf^{0.083} \theta^{0.389} \quad (11.51)$$

$$rSBIS = 2{,}2E - 144 SS\_Y^{380{,}7} Fin\_M^{12{,}91} Sima\_Fin^{8{,}65} Inf^{-1.36} \theta^{-5{,}92} \quad (11.52)$$

$$INF - 1{,}4E - 57\, SS\_Y^{149{,}56} Fin\_M^{14{,}15} Sima\_Fin^{4{,}45} rSBIS^{-0{,}464} \theta^{-3{,}035} \quad (11.53)$$

Tertiary Sector

$$LNTS\_Y = 0.7646 - 0.0228 * LNFIN\_M - 0.0247 * LNSIMA\_FIN$$
$$+0.0025 * RSBIS + 0.0038 * INF + 0.0163 * \theta 1$$
$$t\text{-}stat: (186.1819)***(-2.5758)***(-11.5462)***(24.5238)***$$
$$(14.3573)***(33.5673)***$$
$$R^2 = 0.912237; DW = 0.404019 \quad (11.54)$$

(continued)

(continued)

$$LNFIN\_M = 2.138 - 2.3168 * LNTS\_Y - 0.1622 * LNSIMA\_FIN + 0.0112 * RSBIS$$
$$+0.0093 * INF + 0.0562 * \theta 1$$
$$t\text{-}stat: (3.1441)***(-2.5758)***(-5.8563)***(4.7581)***(2.1588)**$$
$$(3.7412)***$$
$$R^2 = 0.474624; DW = 0.096971 \quad (11.55)$$

$$LNSIMA\_FIN = 16.7742 - 21.3661 * LNTS\_Y - 1.3791 * LNFIN\_M + 0.0577 * RSBIS$$
$$+0.0824 * INF + 0.3986 * \theta 1$$
$$t\text{-}stat: (12.1860)***(-11.5462)***(-5.8563)***(10.9016)***$$
$$(8.03239)***(14.0453)***$$
$$R^2 = 0.757849; DW = 0.380576 \quad (11.56)$$

$$RSBIS = -251.4946 + 329.0075 * LNTS\_Y + 14.2799 * LNFIN\_M$$
$$+8.6595 * LNSIMA\_FIN - 1.3759 * INF - 5.9434 * \theta 1$$
$$t\text{-}stat: (-24.2346)***(24.5238)***(4.7581)***(10.9016)***$$
$$(-14.9623)***(-37.977)***$$
$$R^2 = 0.949888; DW = 0.399031 \quad (11.57)$$

$$INF = -127.4214 + 168.3883 * LNTS\_Y + 4.0723 * LNFIN\_M + 4.2659 * LNSIMA\_FIN$$
$$-0.4746 * RSBIS - 3.0061 * \theta 1$$
$$t\text{-}stat: (-13.8999)***(14.3573)***(2.1588)***(8.03239)***(-14.9623)***$$
$$(-16.0924)***$$
$$R^2 = 0.698992; DW = 0.788969 \quad (11.58)$$

The Cobb-Douglas form is:

$$TS\_Y = 2{,}148 Fin\_M^{-0{,}0228} Sima\_Fin^{-0{,}024} rSBIS^{0{,}0034} Inf^{0.004} \theta 1^{0.0163} \quad (11.59)$$

$$Fin\_M = 8{,}48 TS\_Y^{-2{,}316} Sima\_Fin^{-0.162} rSBIS^{0{,}011} Inf^{0.009} \theta^{0.056} \quad (11.60)$$

$$Sima\_Fin = 19191922 TS\_Y^{21{,}36} Fin\_M^{-1{,}37} rSBIS^{0{,}057} Inf^{0.082} \theta^{0.398} \quad (11.61)$$

$$rSBIS = 6{,}6E - 110 TS\_Y^{329{,}00} Fin\_M^{14{,}279} Sima\_Fin^{8{,}659} Inf^{-1.375} \theta^{-5{,}943} \quad (11.62)$$

$$INF = 4{,}59E - 56 TS\_Y^{168{,}38} Fin\_M^{4{,}072} Sima\_Fin^{4{,}072} rSBIS^{-0{,}475} \theta^{-3{,}006} \quad (11.63)$$

Wellbeing Function

Primary Sector

$$
\begin{aligned}
&\Theta 1 = -157.4784 + 168.3749 * LNPS\_Y + 2.2742 * LNFIN\_M + 1.5803 * LNSIMA\_FIN \\
&-0.1548 * RSBIS - 0.2276 * INF \\
&t\text{-}stat: (-11.4830)***(12.2111)***(1.7090)*(13.7971)***(-32.5914)*** \\
&(-16.0713)*** \\
&R^2 = 0.984777; DW = 0.422095
\end{aligned}
\tag{11.64}
$$

Secondary Sector

$$
\begin{aligned}
&\Theta 2 = -43.6322 + 49.4925 * LNSS\_Y + 5.0699 * LNFIN\_M + 1.6043 * LNSIMA\_FIN \\
&-0.1506 * RSBIS - 0.2276 * INF \\
&t\text{-}stat: (-7.3859)***(6.8668)***(2.92445)***(14.5386)***(-31.3656)*** \\
&(-16.306)*** \\
&R^2 = 0.953019; DW = 0.450415
\end{aligned}
\tag{11.65}
$$

Tertiary Sector

$$
\begin{aligned}
&\Theta 3 = -42.2396 + 55.3807 * LNTS\_Y + 1.8738 * LNFIN\_M + 1.5646 * LNSIMA\_FIN \\
&-0.1554 * RSBIS \;\; 0.2279 * INF \\
&t\text{-}stat: (-32.1785)***(33.5673)***(3.7412)***(14.0453)***(-37.977)*** \\
&(-16.0924)*** \\
&R^2 = 0.972389; DW = 0.422267
\end{aligned}
\tag{11.66}
$$

Statistical Analysis: Results

Primary Sector

From the above estimation results, the rank of each endogenous variable is the value of their inter-elasticity. In the primary sector equation without 'θ' parameterization, the magnitude of elasticity of all variables, that is, financing (Fin_M), SIMA, SBIS rate, and inflation is less than one. This means that the response of primary sector change is inelastic to the change in every independent endogenous variable. Among the overall determinants of the primary sector results, financing is the largest contributor relative to the primary sector, compared to other variables that affect negatively.

The financing elasticity of −0.0899 means that every 1 per cent increase in shari'ah financing will decrease primary sector yield by 0.0899 of 1 per cent, and vice versa, ceteris paribus. Meanwhile, the variables of monetary instrument and macro-inflation variable are inelastic and have little effect on the primary sector. The role of Islamic monetary instruments and the rate of Bank Indonesia Syariah Certificates (rSBIS) are negatively influenced by the primary sector variables. That is, if the primary sector declines, it will cause rSBIS to increase and vice versa, ceteris paribus. In fact, one of the objectives of Bank Indonesia intervention regulated by Law no. 3 year 2004 Article 7 with monetary instruments held is for stabilization of rupiah value, including inflation. In the event of inflation, Bank Indonesia will tend to withdraw the money supply. Meanwhile, if we look at the role of rSBIS to inflation, we find that rSBIS is inelastic to inflation. This shows that the transmission mechanism is differentiated between the monetary sector and the real sector, especially noted for the primary sector.

In the estimation results in the secondary sector, the complementary results are found for transportation and trade, showing the secondary sector result is positively influenced significantly by the financing. This is in contrast to the primary sectors that are negatively affected by shari'ah financing. The role of monetary instrument, rSBIS, is inelastic with a relatively small coefficient value on secondary sector outcomes. Although legally monetary instruments are intended for the purpose of stabilizing inflation, the estimates show that rSBIS does not significantly affect inflation. The strong effect of financing in the secondary sector is also shown by the value of the coefficients of secondary sector being greater than one. The significant positive effect is elastic with Islamic bank financing, while correlated negatively in the primary sector. This implies that Islamic bank financing has not been aligned to the industry as otherwise would be expected. This result indicates that there is an imbalance between the primary sector and the secondary sector. It can be caused by three financing factors in the profit orientation of the sectors: (1) murabaha; (2) short term; and (3) high turnover. In the secondary sector however, the positive impact of shari'ah financing in the secondary sector is shown by positive and elastic influence of Interbank mudarabah Investment (SIMA) and rSBIS.

Tertiary Sector

Furthermore, in the tertiary sector, the monetary instrument of rSBIS also has no significant effect on inflation. The increase in tertiary sector output is supported positively by shari'ah financing, although it is inelastic. So it can be said that the effectiveness of shari'ah monetary instrument is relatively low. There is lack of circular causation linkage between the monetary and real sectors, even though linkage is found to exist with the monetary instruments of shari'ah.

Wellbeing Models

Primary Sector

The monetary transmission model for all three sectors of the real economy gives better result by accommodating 'θ' in the circular causation equations. This is seen from the overall R^2-value being higher. The role of knowledge and the learning process of a participatory attitude implied by 'θ' showed more responsiveness of the endogenous dependent variables to changes in almost all the explanatory variables in every model equation.

The estimated forms of the three-sector estimated wellbeing models result in Cobb Douglas forms:

$$\theta = (-157.4784) PS\_Y^{168,37} Fin\_M^{2,27} Sima\_Fin^{1,58} rSBIS^{-0.155} INF-^{0,22} \tag{11.67}$$

$$\theta = (-43.6322) SS\_Y^{49,49} Fin\_M^{5,06} Sima\_Fin^{1,604} rSBIS^{-0.150} INF-^{0,227} \tag{11.68}$$

$$\theta = (-42.2396) TS\_Y^{55,38} Fin\_M^{1,874} Sima\_Fin^{1,564} rSBIS^{-0.155} INF-^{0,22} \tag{11.69}$$

In Cobb Douglas form, the sum of the variable-specific elasticity coefficients of the wellbeing function gives the indicator of economies of scale of wellbeing. This total elasticity value of each variable to wellbeing provides the quantitative measure interpreting whether economic activity is sustainable by complementarities and stable in the long run or not. Overall, the sum of the coefficients in θ - parameterized equations (estimated wellbeing functions) of all sectors shows a positive magnitude, greater than one.

In the monetary transmission mechanism model, it appears that, the pattern of relationship between the monetary sector and the real sector is unclear. Figures 11.2–11.4 show the flow of influence of monetary instrument to SIMA for the three economic sectors. However, the next process cannot be seen in the structure of interaction in banking and then to the real sector.

The real sector activity is shown by the influence of financing and inflation and vice versa. This pathway reinforces the proof that, the role of monetary instruments is not yet effective in driving the real sector. The disjoint relationship between the monetary and real sectors is still visible for all the three sectors, as shown in Figs. 11.2–11.4. This result also indicates a weak level of participation and complementarities among activities in the transmission mechanism system. It takes effort to get a path that shows the direction of circular causation of each activity. The simulation of the circular causation equations should be structured according to the

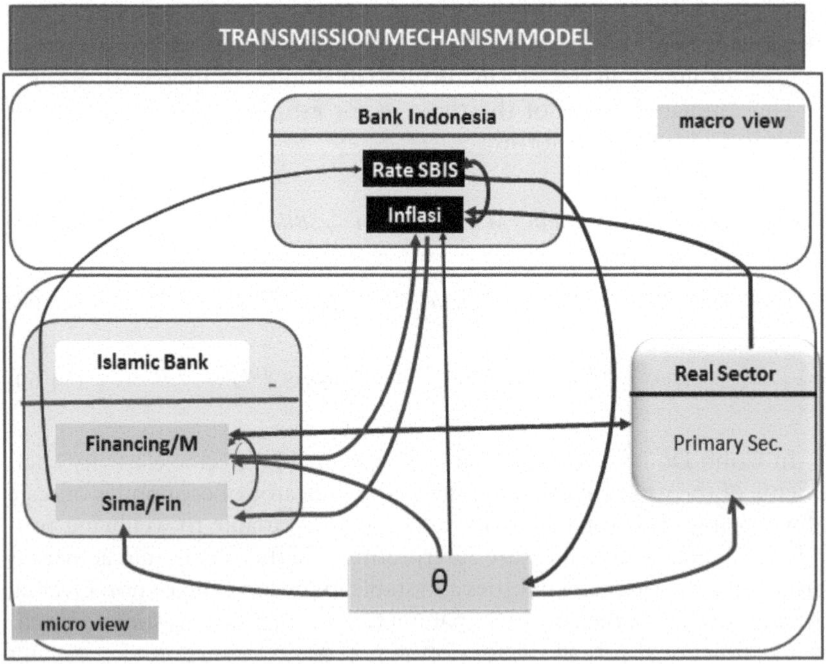

Fig. 11.2 The relationship of the monetary sector to the primary sector

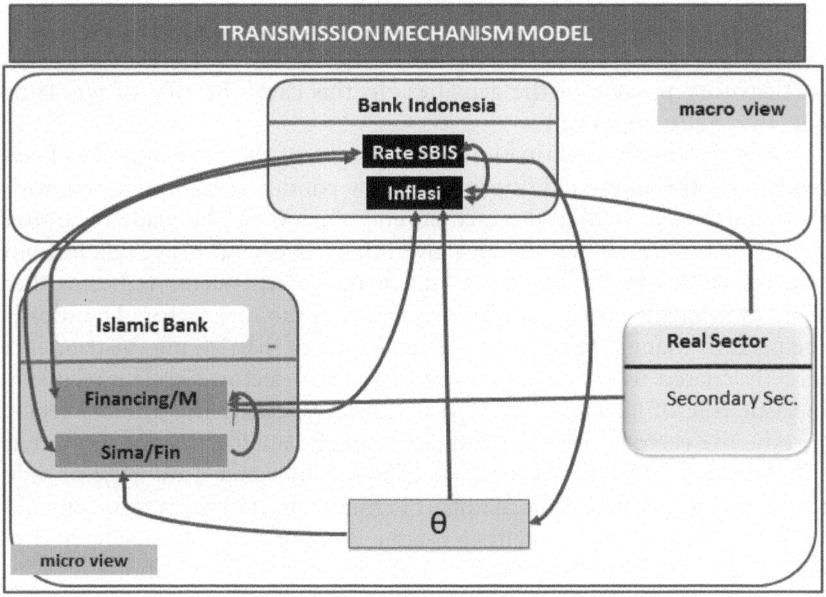

Fig. 11.3 The relationship of the monetary sector to the secondary sector

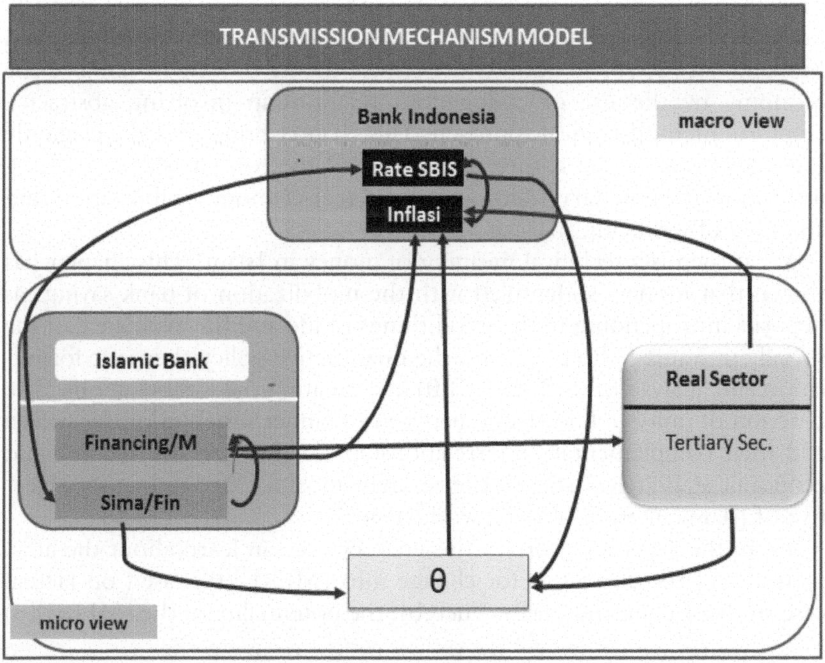

Fig. 11.4 The relationship of the monetary sector to the tertiary sector

participatory process of the activities. In this case, the role of regulator becomes quite important.

For that reason, confirmation of the results of this estimation has been made with the suggested improvement by conducting an interview with the regulators of Bank Indonesia. Several remarks can be made regarding the estimation result and the idea given that (1) unresponsive relationship between rSBIS in shari'ah bank and finally to real sector indicates that shari'ah monetary policy is relatively small to the real sector. The idea is the need to change the monetary instruments of shari'ah into instruments directly related to the development of the real sector. (2) The relatively inelastic role of financing to the real sector, especially the primary sector (productive sector), shows that in fact shari'ah banking has not supported the industry. This happens because Islamic banks take advantage of only three financing instruments, namely (1) murabaha, (2) short term, and (3) high turnover. These are neither Islamic in nature nor effectively diversified to actualize risk-diversification and economic-diversification.

Conclusion

This chapter is among the first of its kind, sparing the contributions by Professor Choudhury (1997, 2015, 2017) and Choudhury, Hossain, and Mohammad (2018) to address the critical theme of how to stabilize the economy by the use of endogenous relationship involving abstracto-empirical methodological approach. This chapter brings out the elegantly brief definition of the nature of money and the role of Islamic banking institution in monetary transmission for real economy stabilization and attaining of wellbeing (maslaha).

What then is a technical meaning of money in Islam? This chapter has shown that money is identified with the mobilization of bank savings as deposits into spending in the good things of life and the avoidance of the socially unwanted choices. Thus, the quantitative policy-theoretic formulation and analysis of the [MFRE](θ) in evaluating the wellbeing objective criterion of unity of knowledge between the inner variables of this model and their complementary interrelationships to theorize and recommend economic stability constitutes the centrally most focused monetary objective of Islamic methodological worldview.

From the empirical work in this chapter, we can learn about the affirmation and the corrective for change into [MFRE](θ) based on factual case study of Bank Indonesia. Thereby, the potentiality of the [MFRE](θ) model and its recommendations in Islamic monetary transmission from Money to Real Economy via the mobilization of Depositor bank savings is established. This abstracto-empirical gateway towards monetary transmission within 100 per cent reserve requirement monetary system transformation comprises a great Islamic transformation and its evolutionary learning centred in Tawhidi methodological worldview.

References

Barrow, J. D. (1991). *Theories of Everything, the Quest for Ultimate Explanation.* Oxford, Eng: Oxford University Press.

Cagan, P. (1989). Monetarism. In J. Eatwell, M. Milgate, & P. Newman (Eds.), *Money, the New Palgrave* (pp. 195–205). New York, NY: W.W. Norton.

Choudhury, M. A. (1997). *Money in Islam.* London, Eng: Routledge.

Choudhury, M. A. (2010). Unity of Knowledge versus Kant's Heteronomy with a Reference to the Problem of Money, Finance and Real Economy Relations in a New Global Financial Architecture. *International Journal of Social Economics, 37*(10), 764–778.

Choudhury, M. A. (2013). *Handbook of Tawhidi Methodology: Especial Reference to Islamic Economics and Finance.* Jakarta, Indonesia: Trisakti University Press.

Choudhury, M. A. (2015). Monetary and Fiscal (Spending) Complementarities to Attain Socioeconomic Sustainability. *ACRN Journal of Finance and Risk Perspectives, Special Issue of Social and Sustainable Finance, 4*(3), 63–80.

Choudhury, M. A. (2017). The Future of Monetary Reform and the Real Economy: An Islamic Problem of Exchange versus Interest. *ACRN Journal of Finance and Risk Perspectives, Special Issue of Social and Sustainable Finance, 6*(4), 37–52.

Choudhury, M. A., & Hoque, M. N. (2017). Comparative Islamic Perspectives in Money, Monetary Policy, and Social Wellbeing. *Journal of Economic Cooperation and Development, 39*(1), 143–162.

Choudhury, M. A., & Hoque, M. N. (2018). Micro-money, Finance and Real Economy Interrelationship in the Framework of Islamic Ontology of Unity of Knowledge and the World-System of Social Economy. *International Journal of Social Economics, 45*(2), 445–462.

Choudhury, M. A., & Rahim, H. A. (2016). An Epistemic Definition of Islamic Economics. *ACRN Oxford Journal of Finance and Risk Perspectives, 5*(2), 106–120.

Choudhury, M. A., Hossain, M. S., & Mohammad, M. T. (2018). Islamic Finance Instruments for Promoting Long-Run Investment in the Light of the Wellbeing Criterion (Maslaha). *Journal of Islamic Accounting and Business Research.*

Choudhury, M. A., Pratiwi, A., Hossain, M. S., & Adenan, F. (forthcoming). In T. Azid (Ed.), *A Relational Wellbeing (Maslaha) Index of Gender Development in Socioeconomic Development Sustainability*. London, Eng: Routledge.

Friedman, M. (1953). Monetary and Fiscal Framework for Economic Stability. In *Essays in Positive Economics* (pp. 133–156). Chicago, IL: University of Chicago Press.

Friedman, M. (1960). *A Program for Monetary Stability*. New York, NY: Fordham University Press.

Hayek, F. A. (1990, Reprint). The Use of Knowledge in Society. In M. C. Spechler (Ed.), *Perspectives in Economic Thought* (pp. 183–200). New York, NY: McGraw-Hill.

Hayek, F. A. (1999). Intertemporal Price Equilibrium and Movements in the Value of Money. In *Good Money, Part One: The New World* (pp. 186–227). Chicago, IL: University of Chicago Press.

Heilbroner, R. L. (1986). The Ideology of Capital. In *The Nature and Logic of Capitalism*, Chapter 5. New York, NY: W.W. Norton.

Mankiw, G. (2003). *Macroeconomics* (5th ed.). New York, NY: Worth Publishers.

Nirdukita, R. (2018). *Monetary-Real Sector Integration Based on Tawhidi String Relation (TSR) Methodology*. On-going PhD Dissertation, IEF-Faculty of Economics, Trisakti University, Jakarta, Indonesia.

Schumpeter, J. A. (1961). *The Theory of Economic Development*. Cambridge, MA: Harvard University Press.

Spencer, N. (2000). On the Significance of Distinguishing Ontology and Epistemology. Retrieved February 2018, from https://ethicalpolitics.org/seminars/neville.htm

Yeager, L. B. (1997). *The Fluttering Veil, Essays on Monetary Disequilibrium*. Indianapolis, IN: The Liberty Press.

CHAPTER 12

Conclusion with an Extension to Tawhid and Shari'ah in Islamic Banking Paradigm

Summarizing the Theme of Tawhid as Foundational Ontological Law

The understanding of the foundational ontological law in Islam, namely the monotheistic oneness of Allah (termed 'Tawhid') as one that meets universal wellbeing in the global sense, rather than being confined to Muslims alone, has been grossly ignored. Academic research concerning the so-called Islamic law, the shari'ah, and the three elements of its inner configuration, namely the Qur'an, Prophetic lessons called the sunnah, and human juristic interpretation and practice known as usul al-fiqh, have been intertwined in an incoherent way. Instead of the truly learned authorities in Islam with deeply exegetic and methodological insight (ulul amr), sheer traditionalists and novices had succeeded. Thus, inept variant interpretations and destructive innovations intruded the highest altar of the Qur'an, sunnah, and found distinctive methodological consequences in the entirety of the creative world-system.

In early days of Islam,[1] shari'ah scholars[2] identified five main securities to mankind around which the spiritual teaching of Islam leads to

[1] The period starts from the time of the Prophet Muhammad (570–632 CE) including the age of the Orthodox Caliphate (632–661 CE), the noble companions, and the succeeding generations till the Umayyad and Abbasid eras (661–1258 CE).

[2] Shari'ah is considered as the sacred law of Islam. It is derived from the Qur'an, the sunnah (the sayings and deeds of the Prophet Muhammad during his lifetime), Ijma' (consensus), qiyas (reasoning by analogy), and maslaha (consideration of the public good or common

M. A. Choudhury, *Tawhid and Shari'ah*,
https://doi.org/10.1007/978-3-030-49087-4_12

prosperity: faith (deen), life, intellect, lineage or posterity, and property rights, which are considered as a part of fiqh al-muamalat. The fiqh as juristic interpretation to justice in all areas of human endeavour has historically served the interests of sects, as well as the mainly financial and legal groups among Muslims. Thus, the exegesis of Qur'anic law, directed by the sunnah—a carrier of the qur'anic law to the generalized and particular cases of the issues and problems under study—has been marred by wishful human interpretations. Examples of such cases are found in the contrariness of interpretations in the shari'ah among different Muslim sects, for example in a broader sense of sunni and shia. To the extent the principal sunni sects further consist of the Hanafi, Shafei, and Hanbali schools of Islamic jurisprudence, which bring complex horizontal and vertical divisions to the communities in Muslim majority countries.

The consequences that ensue as the universal law of Tawhid along with its rich methodological worldview and analytical details have been lost. The principle law of the Qur'an as one that is applicable to Muslims and non-Muslims alike, and is uniquely relevant to the realms of the animate and the cosmological world-systems through abstraction and application, needs to be understood once again. This central law of monotheism (referred to as Tawhid in the Qur'an) has been replaced with a distorted picture of what is now defined as the shari'ah. Shari'ah has turned into a product of misinterpretations by opportunist clerical and theological differences. A great divide thus exists between the definitions of Tawhid as the universal and unique law generality and particulars to all world-systems and shari'ah. This cleavage has happened as a result of the vagaries of human desire, geopolitical, ethno-religious, and commercial interests. An example in this regard is the legitimation of the shari'ah commercial instruments as contracts ('aqd'), contrary to the unified understanding of financing instruments under the Tawhidi methodology, one which integrates and simulates possibilities across unifying systems of thought—the transdisciplinarity. Furthermore, shari'ah has remained in its narrow pale of legal contracts relating to narrowed down worldly affairs, in contrast to a greater domain, which studies worldly matters as well as the abstractions and evidences of universal categories in general (e.g. cosmological entities) (Choudhury, 2015).

need). Yet in this book this interpretation of shari'ah is debated in the light of Tawhid as the sole and ultimate law.

Summary Outlook of This Work

This manuscript has attempted to tackle this complex subdivisional phenomena among human society and introduces the Tawhidi methodological worldview (monotheism) unified model of ontological law of unity of knowledge invoked by the Qur'an and the sunnah. These together enhance spirituality and material prosperity.[3] Thereby, no standardization and harmonization has been determined by a mistaken approach to the identification of what truly is the primal ontological Islamic law. Is it the law of the shari'ah or the universal law of primal ontological origin of Tawhid in the Qur'an, the First Law? Tawhid, that is, the Qur'an is transmitted by the authentic sunnah into the knowledge embedded in 'everything' of the universal world-system.[4] Indeed, here is what Buchman (1998, p. xviii) writes regarding the prevalence of the Tawhidi worldview in the medieval Islamic age being different from those who have failed to understand the res cogitans and res extensa of the Tawhidi weltanschauung: "It (the book, Ghazali's *Niche of Lights*) may be helpful to elucidate the general Tawhid-centered worldview of twelfth-century Islam. People today—Muslims and non-Muslims alike—usually hold drastically different assumptions on the nature of existence than those held by al-Ghazali and most of his contemporaries."

On the topic of development indexing, the shari'ah focuses on 'uqud', separable legal contracts. The Islamic attempts to formulate such development indexing have missed the treatment of the systemic relational organism that is essential to be derived from the Tawhidi unity of knowledge. The consilience of being and becoming (Prigogine, 1980; Wilson, 1998) is a physical model as prototype of unity of knowledge. The algorithmic but not the substantive meaning of Tawhidi methodology of unity of knowledge is formally represented by the consilience model. Yet more significantly, using these analytical models, the Tawhidi methodological worldview of pervasive unity of knowledge with moral induction gives the scheme and order of all unified developmental entities. These are not comprehended either by the idea of biological consilience or the social approach of shari'ah.

[3] *Qur'an* (59: 7): *"So that it (wealth) may not (make) a circuit between the rich among you".*

[4] *Qur'an* (53: 1–3): "By the star when it descends, Your companion [Muhammad] has not strayed, nor has he erred, Nor does he speak from [his own] inclination."

What the shari'ah approach has tried to formulate are humanly self-assigned coefficient values to variables of the development index. A deconstruction of the aggregate index of wellbeing is a principal example as being governed by the purpose and objective of the shari'ah termed as maqasid al-shari'ah. This formulation has been tried by shari'ah scholars but the limits of deconstruction face the problem of differentiated categories in the indexes, indicators, and variables and their inter-causal relations. Thereby, the exogenously assigned numerical values of the coefficients of evaluated models of wellbeing make the whole indexing exercise to be a personated design in existing practices rather than a naturally endowed objective criterion (Davies & Brady, 2015).

Pertinent to such problems is the structure of the linear models of shari'ah scholars. Most often linear forms are used but without a basis of such derivations. At the end, the shari'ah wellbeing model has remained void of formulating the endogenous embedding of moral and ethical values of the Qur'anic genre in the representative variables of the material goods, services, and cognition by ways of abstraction and applications. It is only on the selective common grounds of Tawhid as law in its holistic entirety and in the narrow subset of rules of maqasid al-shari'ah where some commonality can be attained. Nonetheless, this commonality must be accomplished on methodological grounds as endowed by the primal ontology of unity of knowledge and its embedding in the generality and particulars of the world-system. Although the shari'ah continued its intrepid presence in the body framework of Islam for long now, yet the scholarly pursuit for its reconstruction remains alive. This manuscript is meant to be a contribution in such scholarly directions.

Example of the Relationship of Islamic Banking to the Contrasting Views of Tawhid as Law and Shari'ah

The background of Islamic essence of human resource development in bank regulators, managers, and employees and the awareness gap in the broad-based public and businesses ought to be sourced to the weakness of the now inactive Islamicization project a la the intricacies of shari'ah.

The overall financial economics of Islamic banking is to be based on the contrasting study of conflict resolution while examining the existing conflict and competition nature of Islamic banks in the midst of its immersion into capitalist neoliberal paradigm. Shari'ah could not reform such capitalist subservience, while no methodological alternative could be formulated to reject the capitalist neoliberal paradigm. Contrarily, in the case of the Tawhidi methodological worldview in a systemic sense of Islamic banking organically interrelating with economic sectors, the organization of participatory behaviour based on economic cooperation, and institutional cooperation, Islamic banking emerges as a socially inclusive field of study.

Economic development to benefit from the operation of Islamic banks ought to unify the fields of cooperation as market and economic interaction and as institutional organizational behaviour. Such an activity in turn remains deeply dependent on the background of Islamic banks as systems and institutions that inject the Islamic financial instrumentation into economic development in the light of the Tawhidi worldview of unity of knowledge and its functioning. Without such interactive and integrative role of the dynamic learning system of Islamic values, evaluated by the goal of wellbeing function as a measure of social inclusiveness in the sense of coordinated development, there cannot be a sensible meaning of Islamicization and of the truly Islamic functioning of Islamic banks in the development context.

The principal goals of Islamic banks in the light of their socio-economic development coordination are to attain the moral, social, and development objectives that generate and sustain development paradigm of inter-causal unity of goals and purpose between economic system, market system, and the true Islamic objectives of Islamic banks as institutions.

In the case of Islamic banks, wherever they are operating globally today, these financial institutions mushroomed as neoliberal, profit-making, and shareholding outlets without being able to actualize the moral, social, and wellbeing objectives. The innovated proclamation of 'shari'ah-compliance' could not remedy this failed stance. The reason for this missing divide of Islamic banks from the true Islamic worldview as derived from Tawhid as law never dawned in any circle. Thereby, Islamicization and learning on the basis of the true Tawhidi worldview of unity of knowledge between activities, systems, and institutions never dawned. Islamic Bank regulators could not understand this missing gap of learning process in the light of Tawhid as law. Therefore, they could not contribute to the socially inclusive nature of Islamic banks as they ought to be.

Great objectives of the Islamic worldview, namely the nature of Islamic banks as morally and socially embedded institutions in socially inclusive activities, such as stakeholding for the common wellbeing, poverty alleviation, integrated development, invoking of the true Tawhidi ontological worldview of unity of knowledge by interaction, integration, and evolutionary learning, could not be learnt in conceptual and practical sense by Islamic banks. Neoliberal practices continued to abide. The financial and social inclusive principles that ought to be of Islamic banks could not prevail either as a great paradigm or in terms of practices and learning standards of bank managers and Bank Indonesia regulators. These are overwhelming problems of the dysfunction of Islamic banks globally.

There is no special advantage in comparing Islamic banking globally with those in Malaysia as exemplar. In-depth critical examination of the ideas and practices of Islamic banks continues to be deeply rooted in capitalist neoliberal paradigm. Thereby, the theory, learning, and practice of Islamic banks as erroneous palliative of intricate shari'ah rules, mainly centring on riba-free financing instruments, arise from the perspectives of shari'ah enforced by the government viewpoints. The institutions of INCIEF, Bank Negara, ISRA, and every other learning organization on shari'ah rules in the light of the Qur'an and sunnah presently do not exist on the Tawhidi methodological worldview. Consequently, the extant of Islamic banks in socially inclusive areas remain subdued, while capitalist practices of financial globalization abound. Within such a milieu, the so-called 'shari'ah-compliance' abounds.

Islamic bank financial instruments are attenuated by an enforced shari'ah to be free of interest rates. This is a debatable issue. Thereby, the impression and trade yields of Islamic financial practices are questionable in the conceptual and practical sense. Examples are of the interest-consequences of financing by murabaha, sukuk, and many other secondary financing instruments. It is therefore not convincing to accept bare shari'ah rules to pronounce what interest-free financing truly means while not knowing how asset valuation is done in the face of intertemporal meaning of interest rate and, contrarily, not knowing what resource mobilization of bank savings means. Shari'ah investigations have not probed into these subtle issues of analytical nature.

The objective goal of Islamic banks is misguided under the faulty attention of their government regulators, managers, and the Islamicization background. Each of these factors has failed to provide the grassroots-oriented socially inclusive and life-fulfilment perspectives to Islamic banking objectives. The result has thus been a less-than-full trust in Islamic banks by the public at large, which sees the Islamic banks as no different from conventional banks in both goals and operations, but with lesser efficiency of business. Especially marked in such transaction costs of Islamic bank operations is the failure of these banks to diversify risk and production while deepening and widening simultaneously the social capital for wellbeing.

The criterion of profit maximization is a key attribute of capitalist finance. It is now imitated by Islamic banks under the guise of shari'ah permissibility. Contrarily, the dynamics of interactive, integrative, and evolutionary learning experiences arising from Tawhid as law remains unknown and untapped in the Qur'anic sense of inter-causal organic unity of systemic details. Only profit-making for effective operations as an objective can be pursued as business reality. Thereby, participatory financing is not to aim for competition. Instead, a great deal of intra-regional and inter-global participatory networking to mobilize resources is needed. In the regime of life-fulfilment development with resource mobilization in it, the Islamic bank's goal takes the outlook of revenue creation while facing normal profit-making. Consequently, profit accumulation as surplus cannot be a compliant goal to pursue. Within its objective and institutional arrangement of life-fulfilment regime of development, Islamic banks ought to simultaneously maintain affordable pricing and social choices in a stakeholding model of wellbeing. Islamic banks everywhere have failed to provide low-cost service charges. Shari'ah rules remain benign on the profit-pricing issues of financial instruments like murabaha and its secondary financing instruments. The pricing mechanism has remained exorbitant in intertemporal asset financing and valuation for owners. Participatory cost-loading has also remained high in the absence of effective risk-and production diversifications. The knowledge of riba-free financing in the intertemporal sense and thereby of trade activity over the longitude of capitalization over time have remained unknown to Islamic banking and its shari'ah and disengaged gurus. Consequently, Islamic banks have always vouched for shareholder models of capital accumulation, profit maximization, and a crass understanding of the truly Islamic methodological worldview of participation in a networked system of unity of sharing resources so as to integrate Islamic banks in the wider organic development milieu.

In the area of Islamic banking, the application and interpretation of shari'ah rules could not access the study of inner conflicts in concepts and organization of Islamic banking in the truly Islamic epistemological sense. The experiment on Islamicization failed to provide this sustainability energy to Islamic banks. Thus, Islamic banking failed to provide both conception and true application of Islamic wellbeing. The general public distrusts on the contrary with the neoliberal nature of Islamic banking and the inappropriate role of the so-called 'shari'ah-compliance' has deepened. The study of such internal inability of Islamic banks should then be followed up by a study of conflict resolution along lines of the reconstruction of Islamic banking system from the side of the various viewpoints in relation to the Tawhidi ontology of unity of knowledge and the holistic development context.

Substantively advanced Islamic ideas underlying Islamic episteme in economics, finance, and banking ought to provide the prelude to the study of Islamic banking by its problems and prospects. This approach must also be grounded on a model of sustainability and conscious application to moral and social inclusiveness of Islamic banking model within the wellbeing perspective of life-fulfilment regime of development. Human resource development ought to be studied accordingly in the context of the above-mentioned goals (Choudhury, 2016).

In the End

In the end, there is the following concluding remark on the irrelevance of provocations on the end of history and the last worldview (Firstpos', Nov. 3, 2018). This book has provided thought along lines of the role that the philosophy of history plays in consolidating layers of sociopolitical forces that at any juncture of a civilization and nation define a historistic claim in future perspectives. The adherents to this accumulating historistic process then make up the gathering forces to claim their staying power of the future, present, and future. The result then is what is being referred to as jingoism as a force imposed from the pedestal of the upper echelons of power and economy. But even the other counter-philosophy of patriotism that is vouched in a global context for the world civilization and its human, economic, and institutional worldview stands on frail grounds. Indeed, all worldviews must be critically examined against the light of a standard that can be acclaimed as universal and unique to all shades of change and transformations without external coercion. Such a worldview can stand as emulation for all people with experiences profoundly conveyed. The perfect existence of such a worldview may not be fully realized, only learnt along evolutionary learning process.

In this regard the Qur'an and in its embodiment of Tawhid as THE LAW—contra shari'ah, which on a scholarly level cannot be claimed as the universal and unique law of anything let alone the staying universality and uniqueness in 'everything'—is the ideal with its projected applications for all peoples and civilizations between affirmation of historical consciousness and mistaken or learning worldviews. History according to Qur'anic and thus by Tawhidi worldview is an evolutionary learning from and towards that most universal and unique centrepiece of truth of all the worlds, 'everything'. This provocative worldview takes up its indelible mould in the crucible (ontological precept) of oneness of Allah for all of mankind. That worldview is the beginning and end of the primal ontology as law and historical consciousness of unity of knowledge and its induction in 'everything', the world-system.

The Prophet Muhammad's (PBUH) mission was to convey this ultimate message to convey to the world-system of 'everything'. From this final height, many shall learn, act, and perpetuate sustainability over the timeless inter-causal holism of knowledge of unity of being and becoming in knowledge-induced space and time universe in generality and details. No one particular race, people, and civilization, be this Muslim or any other, has the hegemonic presence in this. We all learn in the evolutionary planks of conscious historicism (khalq in-jadid) from the Beginning (Al-Awwal) to the End (Al-Akhir). The world-system of 'everything', comprising civilization and human and related architecture cannot be affirmed in historical consciousness as the ultimate attainment of any singular civilization in its confluence of evolutionary learning premised on the primal ontology of unity of knowledge. This is the domain of Tawhid as THE LAW for everyone regarding 'everything'.

Muslims over history have committed great mistakes, and the end of such overweening jingoism of claims is not seen ending soon. That is the Muslim departure from the essential law of Tawhid and 'everything', and entering the demeaning historical power structure of geopolitics, power, and methodological individualism—contra to the utterance of shari'ah. This is clear from the long-standing behaviour of all people at large, including the clamour of Muslims for something that is contrary to the essence of the Qur'anic, that is, the Tawhidi worldview. The Qur'anic meaning of shari'ah remains different in respect of Tawhid as law (Qur'an, 45:18).[5]

Tawhid as the ontological Beginning and End of the human future is the universal and unique architecture for all, Muslims and other. That is because of the Muslim departure in historical consciousness of the law that was conveyed by the Qur'an through the Prophet Muhammad (PBUH) and all the prophets (anbiyah) to all mankind (An-Nas). In the midst of this overarching of history, Jingoism and patriotic claims disappear while bestowing the greatest gift of Tawhid as law for the entire world-system. The other evanescent claims are replaced by universality and methodological uniqueness. Belief and faith are indeed important manifestations of civilization. Yet these are unseen, unrealized forces inhering in self for inner confirmations. We cannot deal with these attributes in the building of nations, civilizations, society, and world-system. They defy inter-system quantitative aggregation. Only knowledge in its relational organic unity of the good and mutually discoursed choices is of the constructive essence of moral wellbeing (Ghazali). So be it for the global intellectual pursuit.

Within the pail of the above-mentioned nature of historical consciousness of Tawhid as unity of knowledge, the meaning of self and other (dhimmi), meaning the protected minor in religious belief in an Islamic state, is a Qur'anic civilizational reality. It acts as social concept that is coterminous with the majority population expression of civilization as a discursive learning relationship involving 'everything' between self and other in the interactive, integrative, and evolutionary way of process.

[5] Qur'an (45: 18): "Then We have put you on a way of commandment. (So follow you that.) 'Follow what was revealed to you from your Lord, O Muhammad, there is no true deity except Him, and turn away from the idolaters.' Allah said, (and follow not the desires of those who know not)." On this verse, Yusuf Ali explains in the following words: "Shariat is best translated the 'the right way of religion', which is wider than the mere formal rites and legal provisions, which mostly came in the Medina period long after this Meccan verse had been revealed." Ibn Kathir's tafseer of this verse is similar. To the widest meaning on its generality of the verse, "Follow what was revealed to you from your Lord, O. Muhammad"; it is firmly obvious that this specified revelation is the Qur'anic monotheism of Tawhid as law and none other.

References

Buchman, D. (Trans.). (1998). *Al-Ghazali Niche of Lights*. Provo, Utah: Brigham University Press.

Choudhury, M. A. (2015). Res extensa et res cogitans de maqasid as-shari'ah. *International Journal of Law and Management, 57*(6), 662–693.

Choudhury, M. A. (2016). *Islamic Financial Economics and Islamic Banking*. London: Routledge.

Davies, A., & Brady, T. (2015). Explicating the Dynamics of Project Capabilities. *International Journal of Project Management, 9*, 48–58.

Firstpos'. (2018, November 3). *Ramachandra Guha Says jingoism Ascendant due to Hypocrisy of Left, Intellectual, Moral, and Financial Corruption by Gandhis* (Newspaper).

Prigogine, I. (1980). *From Being to Becoming*. San Francisco, CA: W.H. Freeman.

Wilson, E. O. (1998). *Consilience, Unity of Knowledge*. New York: Vantage Press.

Index[1]

[1] Note: Page numbers followed by 'n' refer to notes.

M. A. Choudhury, *Tawhid and Shari'ah*,
https://doi.org/10.1007/978-3-030-49087-4

The manufacturer's authorised representative in the EU is Springer
Nature Customer Service Centre GmbH, Europaplatz 3, 69115 Heidelberg,
Germany. If you have any concerns regarding our products, please
contact ProductSafety@springernature.com

Printed and bound by CPI Group (UK) Ltd, Croydon, CR0 4YY
30/09/2026
02225649-0005